PLANT BIOTECHNOLOGY and DEVELOPMENT

A CRC Series of
Current Topics in Plant Molecular Biology

Peter M. Gresshoff, Editor

Peter M. Gresshoff, PLANT BIOTECHNOLOGY AND DEVELOPMENT, 1992

Forthcoming Titles

PLANT BIOTECHNOLOGY and DEVELOPMENT

Editor

Peter M. Gresshoff
Plant Molecular Genetics
Institute of Agriculture
Center for Legume Research
The University of Tennessee
Knoxville, Tennessee

CRC Press
Boca Raton   Ann Arbor   London   Tokyo

Library of Congress Cataloging-in-Publication Data

Catalog record is available from the Library of Congress.

This book represents information obtained from authentic and highly regarded sources. Reprinted material is quoted with permission, and sources are indicated. A wide variety of references are listed. Every reasonable effort has been made to give reliable data and information, but the authors and the publisher cannot assume responsibility for the validity of all materials or for the consequences of their use.

Direct all inquiries to CRC Press, Inc., 2000 Corporate Blvd., N.W., Boca Raton, Florida, 33431.

© 1992 by CRC Press, Inc.

International Standard Book Number 0-8493-8261-0

Printed in the United States   1 2 3 4 5 6 7 8 9 0

THE EDITOR

Peter Michael Gresshoff, Ph.D., D.Sc., holds the endowed Racheff Chair of Excellence in Plant Molecular Genetics (Institute of Agriculture) at the University of Tennessee in Knoxville.

Dr. Gresshoff, a native of Berlin (Germany), graduated in genetics/biochemistry from the University of Alberta in Edmonton in 1970 and then undertook his postgraduate studies at the Australian National University in Canberra, Australia, where he obtained his Ph.D. in 1973 and his D.Sc. in 1989.

He completed his postdoctoral work as an Alexander von Humboldt Fellow (1973 to 1975) at the University of Hohenheim (F. R. G.) and Research Fellow (1975 to 1979) in the Genetics Department (R. S. B. S.), headed by Professor William Hayes. He was appointed Senior Lecturer of Genetics in the Botany Department at the Australian National University in 1979, where he built up an internationally known research group investigating the genetics of symbiotic nitrogen fixation. He assumed his present position in January 1988, continuing the research direction by focussing on the macro- and micromolecular changes involved in nodulation. His recent interests have turned to plant genome analysis and DNA fingerprinting.

Dr. Gresshoff is also actively involved in the organization of research conferences. In 1990 he chaired the 8th International Congress of Nitrogen Fixation in Knoxville. He also organizes the Annual Gatlinburg Symposia, held in the Smoky Mountains of Tennessee. He is the assistant director of the Center for Legume Research at the University of Tennessee.

He was awarded the Alexander von Humboldt Fellowship twice (1973 to 1975 and 1985 to 1986) and is a member of the editorial board of the Journal of Plant Physiology. He has received major research grants from biotechnology firms and the Australian federal government. He has published over 140 refereed publications and has contributed to many international congresses and symposia. He has been awarded membership in Phi Kappa Phi and Sigma Xi. He is a dedicated teacher and researcher, who believes in technology transfer and innovative science. Presently he is involved not only with federal and international agencies, but also in the commercialization of DAF technology through the establishments of a diagnostic company called GENOMA, Inc. His current major research interest concerns the characterization of the soybean gene controlling supernodulation and nonnodulation as well as the molecules that control these genes.

Table of Contents

Preface

We are experiencing a biological science revolution stemming from the development of subjects that lie between the established scientific disciplines of chemistry, genetics, botany, physics, ecology, zoology, and microbiology. Examples are numerous and obvious. One such new discipline is the field of plant molecular biology, which seeks to understand the molecular chemical mechanisms underlying plant development and plant function. The field naturally is newly emerging and multi-faceted and can be classified anywhere between the purely scientific to the practical and applied. Plant research no longer is fueled only by the desire to increase the agricultural potential of the organism; we are researching plants to understand about the laws of nature itself.

For this reason we are witnessing an expansion in the employment opportunities in this area, as industry and academia recognize the importance of plants. Universities likewise have responded to the new demands as one finds a proliferation of interactive biology departments, schools of biological sciences, and institutes advancing the interdisciplinary approaches of plant molecular biology.

The recent International Congress on Plant Molecular Biology in Tucson bears witness to the immense involvement of all types of biological scientists. The fact that researchers are able to leave established scientific disciplines like yeast or Drosophila genetics and successfully turn their attention to plant development speaks for the fluidity of ideas, techniques, and concepts, mediated by the unifying assumptions of molecular genetics and biochemistry.

The advance of plant molecular biology is now solidly assured through the fundamental discoveries of plant transformation, plant gene isolation, and plant gene analysis in development.

The theme of this series of books is encapsulated in the series title. Current topics are to be covered. These books are not complete textbooks, just like they are not research papers. Each volume of the series will focus on a new and relevant topic that is in a state of flux at the time of its publication. An additional theme of the series is to provide an educational tool. Graduate students and advanced undergraduates require perspective and review. Journal publications are often difficult to understand for the novice or unfamiliar biologist. *"Current Topics in Plant Molecular Biology"* presents a small selection of monographs focused on a particular topic. The authors were given the instruction to be general, while maintaining accuracy and specificity. The mode of publication has stressed the element of speed to allow the reader an up-to-date insight into the subject area.

The first book in the series is entitled ***Plant Biotechnology and Development***. The subject matter is diverse and deals with several biological systems and experimental approaches. Emphasis is given to the area of plant-microbe interactions, especially as illustrated by the legume root nodule symbiosis. The book, therefore, reflects some of the preferences of the editor. Future volumes will be different.

The chapters of this first volume deal with biochemical as well as genetic analysis. Attention was given to morphological as well as evolutionary considerations. A special insight into the perspective from a biotechnology company was given by including a CEO as an author.

Plant Biotechnology and Development is not the complete palette covering the subject. For example, there is no chapter dealing with the new model plant *Arabidopsis thaliana*. There is no mention of flower, embryo, or seed development. I hope that the reader forgives these absences, as these subjects are so frequently and so expertly reviewed in other places.

The reader is also reminded that this is but the beginning. The second volume in this series is being planned (due late 1992) and will be entitled *Plant Responses to the Environment*. We expect an additional volume (due 1994) to deal with plant genome analysis. You can see that there will be ample time to cover organisms and approaches absent from this volume.

Above all, the purpose of these books is to share the enthusiasm of many researchers in plant molecular biology with a new generation of students and to show them that plant biology has careers and rewards similar to those found in more visible disciplines such as medicine and law.

Finally, I want to acknowledge the assistance of The Center for Legume Research at the University of Tennessee, which brought the authors of these monographs together for a small meeting in the mountains of Tennessee. I also want to thank Dr. Brant Bassam for extensive work involving formatting on the Macintosh as well as Janice Crockett for general assistance.

Knoxville, Tennessee
November 1991

Peter M. Gresshoff

DNA amplification fingerprinting and its potential application for genome analysis

Brant J. Bassam, Gustavo Caetano-Anollés and
Peter M. Gresshoff

*Plant Molecular Genetics, Institute of Agriculture and Center for Legume
Research, The University of Tennessee, Knoxville, TN 37901-1071, USA*

Introduction

A DNA sample can be characterized for individual identity according to its chemistry or sequence information. This type of DNA analysis has been referred to as 'fingerprinting', 'profiling', 'typing', 'genotyping' or 'identity testing'. DNA fingerprinting is useful for forensic identification, determination of family relationship, genome linkage mapping, antenatal diagnosis, localization of disease loci, determination of genetic variation, molecular archaeology, and epidemiology (Watkins, 1988; Donis-Keller et al, 1987; Landegren et al, 1988).

Other techniques for biological comparison exist which do not rely on DNA based methodologies, including blood group typing and isozyme analysis. However, the direct analysis of DNA has numerous advantages. Since DNA is the medium of heredity, DNA directly reflects the relatedness or phylogeny of the sample material. Because DNA chemistry (methylation, packaging, etc.) varies between cell types and individual cells inherit base pair mutations at every cell division that are themselves stably inherited, even single cells from an individual can ultimately be characterized. Using more recent techniques involving DNA amplification, very small DNA samples isolated from only a few cells can be analyzed. Due to the high stability of DNA, even mummified or fossilized material can be used (Pääbo, 1989; Golenberg et al, 1990).

Usually, variations in the DNA base sequence are exploited to produce a DNA fingerprint which highlights the uniqueness of a particular sample. Often entire genomes must be examined or compared and current methodology focuses on ways to reduce the enormous complexity of the DNA starting material into simple but characteristic patterns that are representative of the sample.

Conventional DNA fingerprinting

The concept of DNA fingerprinting was founded on the observation by Wyman and White (1980) of a polymorphic DNA locus characterized by a number of variable-length restriction fragments. Thus conventional DNA fingerprinting relies on the detection of what is termed restriction fragment length polymorphisms (RFLPs). Variations in fragment lengths occur when mutations create or abolish restriction enzyme sites in the DNA sample. Conventional analyses involve DNA cleavage by restriction enzymes, electrophoresis of resulting fragments, Southern transfer of separated fragments to a membrane support, radioactive labelling of suitable probes, hybridization of probes to membrane-supported fragments and print detection as a banding pattern on X-ray film. Variant banding patterns, or fingerprints, on a X-ray film are the result of polymorphic fragments detected by a particular probe. The probe DNA may hybridize to multiple tandem-repetitive or hypervariable minisatellites (Jeffreys et al, 1985a; Jeffreys et al, 1985b; Vassart et al, 1987) and produce complex fingerprint banding patterns. Alternatively, probes may be locus-specific for individual hypervariable loci (Wong et al, 1986; Nakamura et al, 1987; Wong et al, 1987; Armour et al, 1989) which produce simpler patterns by detecting alleles from single or even multiple loci.

The polymerase chain reaction

The polymerase chain reaction (PCR) has revolutionized the field of molecular biology (Mullis et al, 1987; Saiki et al, 1988; Ochman et al, 1988). PCR is an *in vitro* method for the enzymatic synthesis of specific DNA sequences which uses two oligonucleotide primers of about 20 nucleotides in length that specifically hybridize to opposite strands flanking the region to be synthesized. Using a repetitive series of cycles involving DNA denaturation, primer annealing and extension of the annealed primers by DNA polymerase, an exponential amplification of the target DNA region occurs. The specifically amplified fragment has its termini defined by the 5' ends of the primers. Since primer extension products are themselves templates for the next round of synthesis in each PCR cycle, the number of products approximately doubles after each cycle. DNA regions of interest can be amplified many million-fold in this way.

The role of PCR in DNA fingerprinting

PCR is useful in conventional DNA fingerprinting for the amplification and incorporation of radiolabeled nucleotides into probes. Probes produced in this way have a very high specific activity which is essential when the ratio of probe to target size is small.

Certain loci, such as the human apolipoprotein B (*apoB*) gene, contain hypervariable regions consisting of a variable number of tandemly repeated short AT-rich sequences (VNTRs). Using tailored primers, the PCR process can specifically amplify such regions (Jeffreys et al, 1988; Horn et al 1990; Boerwinkle et al, 1989). From a sample size of 125 individuals, 12 alleles of the *apoB* gene were identified in this way (Boerwinkle et al, 1989) indicating this procedure has potential for DNA typing. Li et al, (1990) have extrapolated this technique to determine the allelic state of individual sperm cells.

Short repetitive 'simple sequence' motifs such as the stretches of CAG repeats interspersed with CAA triplets in the region of the *Notch* gene of *Drosophila*, are widespread in many genomes. These motifs are generally hypervariable in length. Tautz (1989) exploited this fact in a PCR-based assay that detects such simple sequence length polymorphisms (SSLPs). However PCR artifacts occur which are presently unavoidable.

Jeffreys et al, (1990) used cocktails of specific primers to amplify hypervariable minisatellites from human genomic DNA to produce simple human DNA fingerprints. Amplified products were further characterized by internal restriction enzyme analysis. On occasion however, some minisatellites fail to amplify or produce spurious products.

Orita et al. (1989a) reported a method that detects sequence changes and single-base substitutions. Such sequence variations induce DNA conformation changes, which can be detected as shifts in electrophoretic mobility. They later demonstrated these single-strand conformation polymorphisms (SSCPs) could be detected in PCR amplified and labeled regions (Orita et al, 1989b). Allelic polymorphism in *Alu* repeats at several chromosomal loci were identified using this PCR-SSCP technology.

In all these studies, detection of DNA polymorphisms required considerable experimental manipulation and prior knowledge of DNA sequence or cloned and characterized probe material.

DNA fingerprinting using arbitrary primers

In several recent and independent studies, DNA amplification was directed by single and arbitrary primers to circumvent the requirement for prior DNA sequence information.

Welsh and McClelland (1990) generated reproducible fingerprints of *Staphyloccus* and *Streptococcus* bacteria and three varieties of rice (*Oriza sativa*) using DNA amplification directed by single arbitrary primers. In this study long arbitrary primers were used (20 and 34 nucleotides in length) with two cycles of low stringency amplification followed by high stringency PCR. When separated by polyacrylamide gel electrophoresis and visualized by autoradiography, only about 3 to 20 products predominated. Williams et al (1990) used shorter arbitrary primers 9 or 10 nucleotides in length and 50-80% GC composition, to amplify DNA polymorphisms from several eukaryotic and prokaryotic organisms. They used low stringency cycles and detected up to about 10 products with each primer using agarose gel electrophoresis and ethidium bromide staining. Polymorphic products were used as genetic markers and several of them placed on a soybean RFLP linkage map.

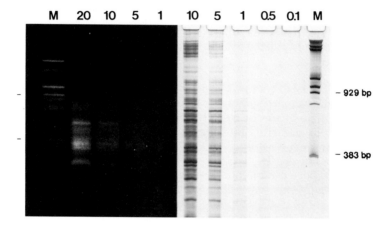

Figure 1. Comparison of DNA separation and staining techniques for DNA fingerprint analysis. Amplification was done in a 100 µl reaction volume with 10 pg of DNA from buffalograss (Buchloe dactyloides cv. Prairie), 1 µg of primer GCCCGCCC (synthesized with > 99% efficiency and HPLC purified to > 95%), 2.5 units of DNA polymerase (AmpliTaq, Perkin-Elmer/Cetus), in reaction buffer [10 mM Tris.HCl (pH 8.3), 50 mM KCl and 2.5 mM MgCl₂] containing 200 µM of each dNTP (Pharmacia). The reaction mix was overlaid with 2 drops of mineral oil, incubated for 5 min at 95 °C, and amplified in an Ericomp thermocycler for 35 cycles (1 sec at 96 °C, 10 sec at 30 °C, and 10 sec at 72 °C, with heating and cooling rates of 23 °C·min⁻¹ and 14 °C·min⁻¹ respectively). Amplification products were separated and stained by (A) agarose gel electrophoresis and ethidium bromide or (B) polyacrylamide gel electrophoresis combined with silver staining (Bassam et al, 1991). The post-amplification reaction mixture was loaded in decreasing amounts (in µl). M; molecular weigh marker.

The relatively simple fingerprint patterns produced in these studies resulted from inherent procedural limitations. This allowed only a few predominant DNA products to be detected. However, complex DNA fingerprints can be resolved using similar methodology when polyacrylamide gel electrophoresis is combined with highly sensitive DNA silver staining (Bassam et al, 1991). Fig. 1 illustrates the importance of using this high-resolution separation and staining technique for DNA analysis. Using this detection procedure, Caetano-Anollés et al (1991) showed that DNA amplification with arbitrary primers as short as 5 nucleotides in length can produce detailed and relatively complex DNA profiles and detect genetic differences when applied to a wide variety of organisms. This procedure has been termed DNA amplification fingerprinting.

DNA amplification fingerprinting

DNA amplification fingerprinting (DAF) uses a PCR-based strategy to amplify short stretches of DNA from a target genome with arbitrarily chosen primers (Caetano-Anollés et al 1991). Amplification under low stringency conditions using a thermostable DNA polymerase directed by one or more short oligonucleotides generates a reproducible spectrum of products. When visualized with an appropriate detection procedure, the spectrum of products resolves into a characteristic fingerprint pattern. DAF uses low stringency amplification conditions to allow primers to anneal at multiple sites on each DNA strand. Although initiation of DNA synthesis occurs throughout the template, only those sequences in which priming sites are on opposite strands and in near proximity will be successfully amplified.

In conventional DNA fingerprinting, polymorphisms result from base pair changes that alter restriction endonuclease sites within defined loci. Similarly, DAF polymorphisms result from changes in DNA sequence within arbitrary primer-defined sites. These changes are manifest in the number and length of products that are successfully amplified and are not linked to particular loci. Several mechanisms can account for these changes. For example, insertion, deletion or inversion of DNA stretches between priming sites can change the size of a particular product or eliminate it entirely. Similarly, base pair substitutions can create and abolish primer sites.

A general observation of DAF fingerprints is that visualized bands fall into two categories; those that are phylogenetically conserved and those that are individual-specific. From this observation it may be concluded that primer sites are randomly distributed along the target genome and flank both conserved and highly variable regions. A second characteristic of all fingerprints examined is wide variation in band intensity. Such variations are reproducible between experiments, however, and could result from multiple copies of the amplified regions in the template or reflect the efficiency with which particular regions are amplified.

Fingerprint tailoring

Fingerprint patterns produced by DAF can be tailored to suit particular requirements. The number of amplification products can be quite variable ranging from none to over a hundred. While product number can be correlated with genome size—larger genomes produce more products—it is largely determined by the primer. In particular, there is some evidence that high primer GC content results in more products. Tailoring the number of products for analysis by selecting appropriate primers can be valuable. Relatively simple patterns are suitable for genetic mapping, while more complex patterns with higher information content are most useful for genotyping. The choice of primer also affects the relative numbers of polymorhic vs. monomorphic products. A higher proportion of polymorphic products is most valuable for determining individual identity whereas monomorphic products are useful for characterization at the species level.

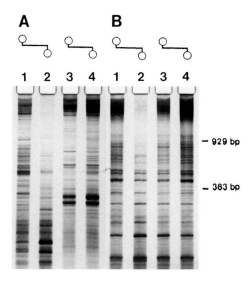

Figure 2. Polymorphic patterns detected in human DNA. Genomic DNA from Caucasian mother and daughter individuals (lanes 1 and 2) and Afro-American mother and daughter individuals (lanes 3 and 4) were amplified with primers of sequence CGCGCCGG (A) or AATGCAGC (B) as described previously (see Fig. 1).

Another way to tailor fingerprints is by using multiple primers, ie. 'Multiplex' DAF. Amplification products obtained using, for instance, two primers produces a fingerprint that is not merely the result of adding the amplification products obtained from each separate primer (Caetano-Anollés et al, 1991).

DAF has been successfully used to detect genetic differences between plants and animals. For example, detailed fingerprint patterns obtained from Caucasian and Afro-American humans can differentiate individuals (Fig. 2). Family related individuals produced patterns that were more monomorphic when compared with unrelated individuals. Again, amplification with certain primers produced conserved patterns among human individuals (Fig. 2). In family studies, amplification fragment length polymorphisms (AFLPs) were inherited from either parent, indicating that these AFLPs can be used as genetic markers in humans (Caetano-Anollés et al, 1991). Different cultivars of the turfgrass *Zoysia japonica* were easily separated and identified with this methodology (Fig. 3). Similarly, AFLPs were detected within the plant genus *Glycine* subgenus *Soja* between the cultivated soybean (*G. max*) and the wild annual species *G. soja*, and between different soybean cultivars (Caetano-Anollés et al, 1991).

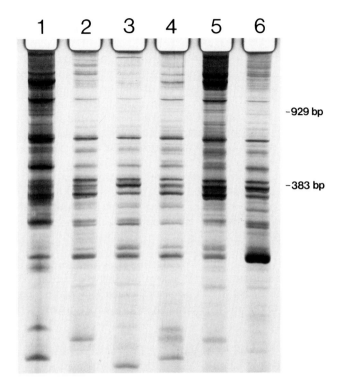

Figure 3. DNA amplification fingerprints of Zoysia japonica cultivars. Amplification profiles were obtained using primer of sequence GCCCGCCC. Zoysia japonica cv. Meyer (lane 1), cv. DALZ-8501 (lane 2), cv. DALZ-8502 (lane 3), cv. DALZ-8507 (lane 4), cv. DALZ-8512 (lane 5), and cv. DALZ-8701 (lane 6). Bands fall into two categories, those that are common to the species and those that in combination are characteristic of a cultivar.

Perspectives

DNA fingerprinting is a powerful tool with largely unexplored potential. Popularity of conventional fingerprinting is hindered by the fact that current methodology is labor intensive, time consuming, requires careful optimization and is, however, subject to experimental variation. It is further limited by the paucity of useful and highly characterized DNA probes. Because DAF is fast and relatively simple, does not require prior knowledge of DNA sequence, is largely independent of the amount of DNA template starting material available and does not require the use and labeling of probes, it represents a conceptual and practical jump over conventional DNA fingerprinting. In a single short experiment DNA fingerprints can be produced with greater resolution and information content than is possible by conventional techniques.

Each of the major procedural steps of DAF is especially amenable to automation. These steps include DNA extraction, robotic manipulation of pre-amplification procedures, amplification, and post-amplification analyses. DAF requires only a small amount of relatively impure template DNA. Thus, the number of enzymatic and chemical treatments currently required for automated DNA extraction procedures can therefore be minimized. Since DAF involves few pre-amplification reaction steps and reagents, automation should be simple and easily linked to the amplification step in the thermocycler. During amplification, the continuous monitoring of products will further optimize the amplification reaction. For post-amplification analysis, we are currently investigating the use of four-color fluorescence detection and computer-based real-time analysis of electrophoretic band patterns (now routinely used for DNA sequencing) to quickly resolve amplification fingerprints with accuracy. The integration and standardization of all these automated steps will be the final objective for the routine fingerprinting analysis of DNA samples.

We believe DAF will be valuable for identity testing, population and pedigree analysis, phylogenetic studies, genetic mapping and molecular characterization of near isogenic lines. The use of very short oligonucleotide primers should be particularly useful for high density mapping of polymorphic markers.

Since amplification of RNA species by way of their cDNA homologues is also possible, we anticipate using DAF to generate fingerprints reflecting gene expression. The high sensitivity of DAF should allow direct identification of cDNA from even weakly expressed genes.

Acknowledgement

The first two authors made equal contributions to this work. Funds were provided through the Racheff Chair of Excellence in Plant Molecular Genetics. We thank L. Callahan for preparing turfgrass DNA samples. Some zoysiagrass cultivars were obtained from M.C. Engelke, Texas A&M University, Dallas. DAF is the subject of two US Patent Applications.

References

Armour, J.A.L., Wong, Z., Wilson, V., Royle, N.J. & Jeffreys, A.J. (1989) *Nucleic Acids Res.* **17**, 4925-4935.

Bassam, B.J., Caetano-Anollés, G. & Gresshoff, P.M. (1991) *Anal. Biochem* **196**, 80-83.

Boerwinkle, E., Xiong, W., Fourest, E. & Chan, L. (1989) *Proc. Natl. Acad. Sci. USA* **86**, 212-216.

Botstein, D., White, R.L., Skolnick, M. & Davis, R.W. (1980) *Am. J. Hum. Genet.* **32**, 314-331.

Caetano-Anollés G., Bassam, B.J. & Gresshoff, P.M. (1991) *Bio/Technology* **9**, 553-557.

Donis-Keller, H., Green, P., Helms, C., Cartinhour, S., Weiffenbach, B., Stephens, K., Keith, T.P., Bowden, D.W., Smith, D.R., Lander, E.S., Botstein, D., Akots, G., Rediker, K.S., Gravius, T., Brown, V.A., Rising, M.B., Parker, C., Powers, J.A., Watt, D.E., Kauffman, E.R., Bricker, A., Phipps, P., Muller-Kahle, H., Fulton, T.R., Ng, S., Schumm, J.W., Braman, J.C., Knowlton, R.G., Barker, D.F., Crooks, S.M., Lincoln, S.E., Daly, M.J. & Abrahamson, J. (1987) *Cell* **51**, 319-337.

Golenberg, E. M., Giannasi, D.E., Clegg, M.T., Smiley, C.J., Durbin, M., Henderson, D. & Zurawski, G. (1990) *Nature*. **344**, 656-658.

Horn, G.T., Richards, B. & Klinger, K.W. (1989) *Nucleic Acids Res.* **17**, 2140.

Jeffreys, A.J., Wilson, V. & Thein, S.L. (1985a) *Nature* **314**, 67-732.

Jeffreys, A. J., Wilson, V. & Thein, S.L. (1985b) *Nature* **316**, 76-79.

Jeffreys, A.J., Wilson, V., Neumann, R. & Keyte, J. (1988) *Nucleic Acids Res.* **16**, 10953-10971.

Jeffreys, A.J., Neumann, R. & Wilson, V. (1990) *Cell* **60**, 473-485.

Landegren, U., Kaiser, R., Caskey, C.T. & Hood, L. (1988) *Science* **242**, 229-237.

Li, H., Xiangfeng, C. & Arnheim, N. (1990) *Proc. Natl. Acad. Sci. USA* **87**, 4580-4584.

Mullis, K.B. & Faloona, F.A. (1987) *Meth. Enzymol.* **255**, 335-350.

Nakamura, Y., Leppert, M., O'Connell, P., Wolff, R., Holm, T., Culver, M., Martin, C., Fujimoto, E., Hoff, M., Kumlin, E. & White, R. (1987) *Science* **235**, 1616-1622.

Ochman, H., Gerber, A.S. & Hartl, D.L. (1988) *Genetics* **120**, 621-623.

Orita, M., Iwahana,H., K., Kanazawa, H,Hayashi, K & Sekiya (1989a) *Proc. Natl. Acad. Sci. USA* **86**, 2766-2770.

Orita, M.,Suzuki, Y., Sekiya, & T.Hayashi, K. (1989b) *Genomics* **5**,874-879.

Pääbo,S (1989) *Proc. Natl. Acad. Sci. USA* **86**, 1939-1943.

Saiki, R.K., Gelfand, D.H., Stoffel, S., Scharf, S.J., Higuchi, R., Horn, G.T. & Erlich, H.A. (1988) *Science* **239**, 487-491.

Tautz, D. (1989) *Nucleic Acids Res.* **17**, 6463-6472.

Vassart, G., Georges, M., Monsieur, R., Brocas, H., Lequarre, A.S. & Cristophe, D. (1987) *Science* **235**, 683-684.

Watkins, P.C. (1988) *Biotechniques* **6**, 310-320.

Welsh, J. & McClelland, M. (1990) *Nucl. Acids Res.* **18**, 7213-7218.

Williams, J.G.K., Kubelik, A.R., Livak, K.J., Rafalski, J.A. & Tingey, S.V. (1990) *Nucleic Acids Res.* **18**, 6531-6535.

Wyman, A.R. & Whyte, R. (1980) *Proc. Natl. Acad. Sci. USA* **77**, 6754-6758.

Wong, Z., Wilson, V., Jeffreys, A.J. & Thien, S.L. (1986) *Nucleic Acids Res.* **14**, 4607-4616.

Wong, Z., Wilson, V., Patel, I., Povey, S. & Jeffreys, A.J. (1987) *Annu. Hum. Genet.* **51**, 269-288.

The plant molybdenum cofactor (Moco) - its biochemical and molecular genetics

Ralf R. Mendel

Institute of Genetics and Crop Plant Research, O-4325 Gatersleben, Germany

Introduction

In this paper I review part of our previous work on the biochemical and molecular genetics of the plant molybdenum cofactor. For a more detailed review see Müller and Mendel (1989).

The moybdenum cofactor (Moco) is a component common to all molybdoenzymes with the exception of nitrogenase. It was shown to be a low-molecular-weight molybdopterin exhibiting no catalytic activity on its own but becoming biologically active on association with an appropriate apoprotein. Nitrate reductase (NR) is the most important plant molybdoenzyme followed by xanthine dehydrogenase (XDH). The Moco-containing molybdoenzymes sulfite oxidase and aldehyde oxidase occur mainly in animals, pyridoxal oxidase in insects and several different molybdo-enzymes are known in bacteria.

The Moco has three general characteristics:

1. *unique* in core structure which is the same in all organisms so far examined,

2. *universal* in function in that it forms part of the catalytically active center of very diverse enzymes thus catalyzing diverse reactions on C-, S- and N-atoms

3. *ubiquitous* in occurrence in that it occurs in all living organisms ranging from bacteria to humans.

Moco mutants have been described in bacteria, fungi, algae, mosses, plants, insects and humans. It is remarkable that in all cases not only one but several mutant loci have been described. So one can assume that the process of Moco synthesis is a multi-step pathway. Whether or not this pathway is similar in all these organisms remains to be seen. The isolation and molecular comparison of Moco genes of organisms of different phylogenetic origin will answer this question. But, at first mutants in theses loci had to be characterized in greater detail.

Biochemical characterization of Moco mutants

In the tobacco species *Nicotiana plumbaginifolia,* mutant plants defective in the molyboenzyme NR have been isolated and studied. Two mutant classes have been found: *nia* mutants (defective in the structural gene of NR) and *cnx* mutants impaired in the Moco. At least six genes (*cnx*A to *cnx*F) are involved in the synthesis of active Moco. A mutation in any of these six genes leads to a simultaneous loss of NR and XDH activities due to a defect in the common Moco. The phenotype of the *cnx* mutants allows conclusions to be drawn also about the physiological role of XDH and possibly other enzymes containing Moco.

A detailed biochemical and genetic analysis of the *cnx* mutants gives us further insight into the localization of the defect and hence contributes to a better understanding of the synthesis of Moco in plants.

The NR apoprotein is not affected by a *cnx* mutation. The obvious complementary nature of *cnx* and *nia* mutants has not only been demonstrated genetically (by somtic hybridization and by crossing of regenerated mutant plants) but was also confirmed biochemically by reconstitution of NR activity in mixed extracts prepared by co-homogenization of *cnx* and *nia* mutant cells. In these experiments the inactive NR aproprotein of *cnx* mutants could be reactivated in vitro by the active Moco present in *nia* mutant cells. All attempts to demonstrate in vitro restoration of NR activity by homogenizing together cells mutated in different *cnx* genes remained unsuccessful. Since *cnx* mutants complement each other genetically but not in vitro after extract mixing, it may be concluded that either the products of the *cnx* genes are very unstable under in vitro conditions of that the synthesis of Moco is linked to processes and/or structures that are rendered non-functional once the cell has been extracted.

Assay for Moco - the nit-1 system

As a next step we had to find parameters and establish methods that would allow a more precise description of the impairment caused by a given *cnx* mutation. In order to assay Moco in vitro a biotest was used since Moco by itself has no catalytic activity. It gains catalytic activity only by combining with an appropriate aproprotein. Therefore extracts of the Moco mutant *nit*-1 of *Neurospora crassa* lacking Moco were incubated with a Moco source, and the activity of restored *Neurospora* NR was taken as a measure of the amount of active Moco.

When the *nit*-1 assay was performed in the presence of a high concentration of molybdate, *cnx*A mutants were shown to contain a cofactor moiety which could complement *nit*-1 to high activity. Omitting molybdate from the assay prevented *nit*-1 complementation, whereas *nia* mutants do complement *nit*-1 in the absence of molybdate. It can be assumed, therefore, that the defect in the *cnx*A mutants resides in a lack of molybdenum as the catalytically active ligand metal for the cofactor, while the structural pteridine moiety of the cofactor does not seem to be impaired by the mutation. Extracts of the other *cnx* mutants do not complement *nit*-1.

Molybdenum content

To rule out the possibility that the *cnx*A mutants are uptake mutants for Mo the cellular Mo content was determined. Mutants in the *cnx*A gene and in all the other *cnx* genes contained Mo in amounts similar to wild-type and *nia* cells and are hence unlikely to be uptake mutants for Mo.

Repair by molybdate in vivo and in vitro

The *cnx*A mutants were shown to contain a potentially active molybdopterin moiety, presumably lacking Moco in situ. Subsequently the cells were grown on media containing high levels of molybdate in attempts to circumvent the role of the gene product impaired by the *cnx* mutations. Of all the *cnx* mutants tested only *cnx*A lines were repairable under these conditions.

When *cnx* cells were extracted in the presence of high concentrations of molybdate together with thiol reagents, NR activity was regained only in the case of *cnx*A. Thus, in the presence of high levels of molybdate the function controlled by the *cnx*A gene is neither required in vivo nor in vitro for the formation of active NR.

Repair in vitro by heterologous Moco

The addition of heterologous Moco prepared from bovine milk xanthine oxidase to cell extracts of *cnx* mutants efficiently restores NR in all types of *cnx* mutants thereby demonstrating both the functional integrity of the *nia*-coded NR apoprotein and the universal function of Moco. Repair by heterologous Moco of animal, bacterial (*E. coli*) and fungal (*N. crassa*) origin, was also possible. The ability of NR apoprotein to act as an acceptor for Moco of different origins seems to be a general characteristic property of *cnx* mutants in *Nicotiana*.

Moco not only forms part of the catalytically active center of NR but is also essential for dimerizing the monomeric subunits of NR. By performing a sucrose density gradient centrifugation of extracts from *cnx* mutants and measuring the diaphorase activity of the subunits, one can check the dimerization state of the NR apoprotein. Only *cnx*A and one of the *cnx*D mutants showed dimeric NR apoproteins; *cnx*B and *cnx*C mutants were monomeric, whilst *cnx*E and other *cnx*D mutants gave intermediate results.

Functions of the *cnx* gene products

*Cnx*A and *cnx*D cells contain a Moco that, although defective in its catalytic properties, is still able to mediate dimerization of NR monomers. The defect caused by the *cnx*D mutation is obviously quite different from that caused by the the *cnx*A mutations since *cnx*D mutant cells cannot be repaired by molybdate in vivo or in vitro and do not complement *nit*-1. Mutations in *cnx*D therefore cause a partial structural defect in the molybdopterin moiety of Moco whereas mutations in *cnx*A leave the whole molybdopterin structurally unaffected.

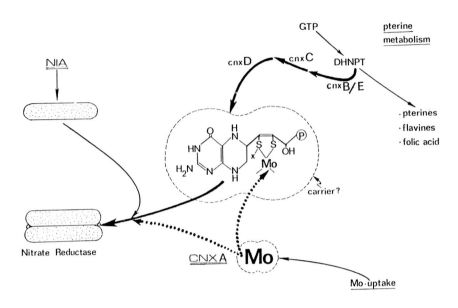

Fig. 1 Working model for the pathway of Moco biosynthesis in N.plumbaginifolia.

It seems reasonable to suggest that the *cnx*A gene product is not involved in the synthesis of molybdopterin but is essential for inserting molybdenum into the pteridine moiety of Moco. Mo insertion seems to be the final step in the biosynthesis of active Moco and is thus the link between two metabolic pathways, one of which involves the biosynthesis of the pteridine moiety of Moco, the other pathway involving uptake, transport and processing (chelation) of the molybdate anion.

There are no currently measurable biochemical differences between *cnx*B, *cnx*C, *cnx*D, *cnx*E and *cnx*F mutants. Therefore further classification parameters have to be developed in order to discriminate biochemically between these loci. Mutants in these loci presumably lack molybdopterin or possess a form so heavily damaged that it cannot be detected by the approaches used, suggesting that theses genes are involved in the synthesis of the molybdopterin moiety of Moco.

A working model for this new metabolic pathway in plants is shown in Fig.1.

Since the pterin ring systems forms also part of other compounds, some of which are essential for cellular metabolism (e.g. folic acid, flavins), a mutational block in its synthesis would be lethal for the cell. Hence the products of these *cnx* genes are more likely to be involved in the insertion of the functional groups, in particular the important sulfur atoms. The existence of Moco carrier/storage proteins should also be taken into consideration since free pteridines are generally of low solubility in aqueous media.

Cloning of plant *cnx* genes

How to clone a plant gene whose gene product you do not know and for which the only fact you have is the phenotype of a mutant? In order to solve this question we planned to use the *cnx* mutants as recipients for transformation experiments using whole libraries to be transferred in the mutants and selecting for the re-appearance of NR activity, i.e. growth on medium containing nitrate as sole N-source. Two approaches were tried (Schiemann et al., 1990):

1. Genetic complementation of a cnx *mutant by transferring a plant genomic library and selecting for cells with restored NR activity*

A cosmid library of *Arabidopsis thaliana* genomic DNA (mean insert size 23.6 kb) in the binary vector pC22 was transformed into *E. coli* K803 and mobilized into *Agrobacterium tumefaciens* C58C1 containing the disarmed Ti-plasmid pGV2275. Plant cells of the *cnx*A-type have been transformed by protoplast co-cultivation as well as by leaf disk transformation. In the latter case, shoots were selected for resistance to kanamycin (*npt*) as genetic marker on the binary vector. The second round of selection was performed by transferring the kanamycin-resistant shoots to nitrate medium in order to select for nitrate utilization.

The experiment was performed two times with the Moco mutant CnxA135 (and also once, for control reasons, with the structural gene mutant Nia30). Although in each case about 100,000 kanamycin-resistant shoots have been raised, not a single shoot was able to survive on nitrate medium. A thorough statistical analysis of this approach (Simoens et al., 1986) revealed that at least 540,000 kanamycin-resistant shoots are necessary to have the chance (95% probability) of finding one *cnx* gene. This number, however, is too high for the leaf disk approach.

In the case of the protoplast co-cultivation, cells were selected at first for nitrate utilization. All NR-positive calli were subsequently transferred to kanamycin medium and were subjected to Southern genomic DNA hybridization. No NR-positive clone turned out to be transformed. Thus only revertants were selected in this way.

These very risky experiments of trying to select a gene by genomic shot-gun cloning on the plant cell level were therefore stopped and another more promising approach was explored.

2. Genetic complementation of E.coli chl *mutants (being defective in the Moco) by a tobacco cDNA expression library, and using of the selected cDNA for transforming tobacco* cnx *mutants as recipients*

Since genomic shot-gun cloning directly on the plant cell level was not successful we tried to circumvent the more complicated eucaryotic level by using procaryotic Moco mutants as recipients for a plant expression library.

A cDNA library (in pUC18) of *Nicotiana tabacum* (tobacco) suspension culture cells was established. The cells were kept in the dark in order to enrich for house-keeping mRNAs being not related to light. The library was transformed into super-competent *chl* mutants of *E. coli* being defective in the Moco (i.e. these *E. coli* mutants are analogous in their phenotype to the plant *cnx* mutants). These mutants, generated by Tn-mutagenesis, were obtained from V. Stewart (USA). In *chl*G reproducible complementation was observed. Restriction analysis of inserts selected in independent experiments revealed always the same type of insert. This 0.8 kb DNA was sequenced and turned out to be not full-length (lacking the 5' terminus). A search in data bases gave no homology to any other sequence.

This cDNA was taken as a probe to isolate the corresponding genomic sequence from a phage library (Charon 4) of *Nicotiana plumbaginifolia*. Three clones were selected and shown to represent a similar insert. The characterization is in progress.

The cDNA was also recloned into a binary plant expression vector (pC27) under control of the strong TR2' promoter. This vector construct was taken for direct gene transfer into protoplasts of *cnx* mutants (electroporation and/or PEG treatment with subsequent test for transient gene expression). This means that 3 to 5 days after transformation the cells were tested for NR activity. *Cnx*A cells were not complementable by the selected cDNA, however cells from *cnx*B as well as *cnx*C, *cnx*D, *cnx*E and *cnx*F were all complementable, i.e. they showed about 20% wild type NR activity whereas the controls with the insert-less vector gave reproducibly no NR activity. One highly speculative explanation could be that we cloned with the incomplete cDNA a domain common to several *cnx*-encoded Moco processing enzymes. However, this speculation is contradicted by the finding that this cDNA sequence taken as a probe for genomic hybridizations did give only one hybridizing band in blots of genomic DNA from tobacco, alfalfa, rice and *Arabidopsis*.

The final proof that the selected sequence is *cnx*-related will come from the stable genetic complementation of a *cnx* locus by the recloned genomic sequence. This work is in progress.

References

Müller, A.J. & Mendel R.R. (1989) Biochemical and somatic cell genetics of nitrate reduction in *Nicotiana*. In: Molecular and Genetic Aspects of Nitrate Assimilation, Wray, J.L. & Kinghorn, J.R. (eds.), Oxford Science Publications, Oxford pp 166 - 185.

Schiemann, J., Inze, D. & Mendel, R. (1990) In: Nitrate assimilation - molecular and genetic aspects, 3rd International Symposium, Bombannes (France), p. 141 - 142

Simoens, C., Alliotte, T., Mendel, R., Müller, A., Schiemann, J., Van Lijsebettens, M., Schell, J., Van Montagu, M. & Inze´, D. (1986) *Nucleic Acid Res.* **14**, 8073-8090.

Motility and chemotaxis in the life of rhizobia

Wolfgang D. Bauer

Departments of Agronomy, Microbiology and Plant Biology, The Ohio State University, Columbus, OH 43210

Introduction

The aim of this brief review is to consider how motility and chemotaxis may contribute to the success of bacteria in natural environments. A great many species of bacteria are both motile and chemotactic. The ability to move, and the ability to detect and swim towards local sources of nutrients would seem to be valuable assets in almost any environment. But motility and chemotaxis require a major commitment on the part of the cell: in *E. coli* for example, over 50 different genes needed for chemotaxis have been identified, and an appreciable fraction of a cell's total resources must be expended to synthesize and drive the resultant machinery (McNab 1987a,b). Moreover, bacteria may encounter many circumstances in their native environments where chemotaxis would be impossible, fruitless or counterproductive: eg. when cells are stuck on surfaces, trapped in pores, or distant from any source of nutrient. Thus, it is not clear when, and how, and how much motility and chemotaxis actually contribute to the success of bacteria under natural conditions. Nor is it clear whether bacteria are able to regulate their swimming and tactic activity in response to environmental circumstances so as to make their investment in chemotaxis proportional to the probability of return. To date, these issues have not really been addressed for any bacterial species in any natural environment (Chet and Mitchell 1976; McNab 1987b).

In trying to understand the functioning of bacterial chemotaxis in natural environments, there is fortunately a good base of knowledge about the molecular-cellular mechanics of chemotactic behavior (Koshland 1981; McNab 1987b). Most motile bacteria swim by means of rotating propellers called flagella. They are able to bias their movement towards higher concentrations of an attractant compound by linking changes in attractant concentration to changes in flagellar rotation. Counterclockwise rotation of the flagella results in smooth forward propulsion, whereas pauses or clockwise rotation result in random reorientation of the bacteria. Binding of an attractant to membrane transducer proteins causes conformational changes which indirectly increase the proportion of time that the flagella rotate CCW. The resultant bias towards longer swims in the presence of attractant creates net movement of the bacteria in the direction of increasing attractant concentration.

Motility and chemotaxis of rhizobia

Rhizobia are root colonizing bacteria capable of living independently in the soil for many years. They also infect and survive in the root cells of host legumes for long periods by stimulating the root cells to form nodules in which the bacteria fix atmospheric nitrogen (Rolfe and Gresshoff 1988). All rhizobia are both motile and chemotactic, capable of responding to a wide diversity of organic compounds.

Mutants of rhizobia which have lost motility and chemotaxis are still capable of establishing a normal symbiosis with its host legume (Napoli and Albersheim 1980). Thus, movement and chemotaxis are not essential parts of the infection process. However, Mot⁻ and Che⁻ mutants are found to occupy only 5% to 30% of the root nodules formed during competition against equal numbers of the wild-type. Thus, defects in motility or chemotaxis significantly impair the symbiotic competitiveness of rhizobia (Ames and Bergman 1980; Mellor et al 1987; Caetano-Anollés et al 1988a; Catlow et al 1990; Liu et al 1989).

The experiments of Liu et al (1989) provide the strongest evidence thus far for a positive role of motility and/or chemotaxis in the life of rhizobia since they were conducted in soils held at realistic moisture levels and populated with indigenous microbes competing for whatever substances serve as nutrients and attractants for rhizobia. Similar evidence that chemotaxis contributes to the competitive success of bacteria in natural environments has been obtained with *Vibrio cholerae* in the intestinal mucosa of its hosts (Allweis et al 1977) and with *Pseudomonas syringae* on bean leaves (Haefele and Lindow 1987).

Liu et al did not attempt to find out how chemotaxis contributed to the competitive advantage of the wild-type over the non-motile mutant, nor did they attempt to determine the relative importance of blind swimming (motility) vs directed swimming (chemotaxis). The observed superiority of their wild-type *Bradyrhizobium* strain could have been a result of enhanced saprophytic survival in the soil, enhanced colonization of the host root, or enhanced efficiency of symbiotic infection. Further studies are clearly needed to distinguish between these possibilities. The role of motility and chemotaxis in root colonization has been examined in several studies, but these have not yet come to any generalizable conclusions (cf Catlow et al 1990; Howie et al 1987; Reynolds et al 1989; Scher et al 1988; De Weger et al 1987).

Rhizobium meliloti has been the base for our recent studies because its chemotactic machinery and behavior are best characterized among rhizobia. Various workers have examined its chemotactic responses to diverse substances (Götz et al 1982; Bergman et al 1988; Caetano-Anolles et al 1988b; Malek 1989; Dharmatilake 1990). Behavioral mutants have been isolated, characterized and mapped (Ames and Bergman 1980, 1981; Ziegler et al 1986; Bergman et al 1988; Ronco 1988). Two independently transcribed flagellin genes have been sequenced (Pleier and Schmitt 1989). Methods have been developed for examining chemotaxis of *R. meliloti* in soils. In the rhizosphere of alfalfa seedlings, videomicroscopy has revealed the formation of chemotactic swarms or "clouds" of *R. meliloti* at a number of very small, discrete sites in the symbiotically infectible region of the root (Gulash et al 1984; Malek 1989). These sites of chemotactic attraction were otherwise indistinguishable from adjacent areas on the root surface. *R. meliloti* formed similar localized swarms on the root surfaces of other legumes but not on non-legumes. Other bacterial genera were not attracted to the same local sites as rhizobia, so the dominant chemoattractants released from these sites seem to be specific to legumes, and for rhizobia. Their chemical identity is not known. Mutants which form chemotactic swarms on the host root surface, yet are defective in responses to ordinary attractants such as amino acids have been isolated. This led to the suggestion that *R. meliloti* may have two independent pathways of chemotactic signal transduction, one for ordinary nutrients and another for certain host-related substances (Bergman et al 1988).

Role of motility and chemotaxis in nodule initiation

In experiments where alfalfa seedlings were immersed in a 1:1 mixture of wild-type *R. meliloti* and Mot⁻ or Che⁻ mutants, the wild-type firmly attached to the infectible zone of the root about 10 times more efficiently than the mutants (Caetano-Anollés et al 1988a). Likewise, when *R. meliloti* cells were dribbled onto the host root surface, the wild-type was found to be, cell for cell, 10 to 30 times more efficient in generating the first nodules than either Mot⁻ or Che⁻ mutant derivatives, at least at low to medium inoculum dosages. Since attachment and initiation of the first nodules occurs within a few hours after contact of the bacteria with the root, it appears that both motility and chemotaxis can contribute significantly to the early stages of symbiotic infection, even before the first round of bacterial multiplication. Wild-type *R. meliloti* appears to have a considerable advantage in nodule initiation even when applied in very thin films to ensure that the non-motile mutant came into equal contact with the root surface (Caetano-Anollés, Wrobel-Boerner and Bauer, unpublished). This suggests that motility and chemotaxis somehow serve to enhance infection efficiency even after initial contact, perhaps via movement over the root surface to microsites especially favorable for infection.

Chemotaxis towards *nod* gene inducing phenolics from the host root

Caetano-Anollés et al (1988b) and Dharmatilake (1990) have shown that *R. meliloti* responds chemotactically to at least two of the major flavonoid inducers of *nod*ulation gene expression: to luteolin, which is present in alfalfa seed exudates, and to 4',7-dihydroxyflavone, which is present in alfalfa root exudates (Maxwell et al 1989). Chemotaxis to these compounds was biochemically very specific. Maximum chemotactic responses to the flavonoids occurred at 10^{-7} to 10^{-9} M, concentrations which are about 10- to 100-fold lower than those required for half-maximal *nod* gene induction. Similar chemotactic responses to host phenolics have been reported for other rhizobia and for *Agrobacterium* (Parke et al 1987; Shaw et al 1988; Aguilar et al 1988; Armitage et al 1988). It is thus possible that host recognition in terms of gene expression is coupled to chemotaxis in many bacteria. However, it is still not clear how much these chemotactic capabilities actually contribute to accumulation of rhizobia in the host rhizosphere and consequent enhancement of *nod* gene induction. The *nod* inducing flavonoids are secreted rapidly enough by alfalfa roots to reach chemotactically active concentrations at distances several millimeters from the surface (Maxwell et al 1989), and in contrast to other root exudate components, the flavonoids are not readily metabolized by microbes. On the other hand, flavonoids are rapidly adsorbed to surfaces and the chemotactic responses they elicit appear to be relatively weak, at least with laboratory cultured bacteria. Moreover, *R. meliloti* can spontaneously lose its chemotactic responsiveness to the *nod* gene inducers, even though it remains responsive to other attractants (Dharmatilake 1990).

Changes in behavioral activity in response to nutrient limitation

Most bacteria, including rhizobia, live most of their lives under constant pressure of nutrient insufficiency. The competitive success of bacterial species in nutrient-poor environments such as soil thus depends to a large extent on their specific adaptations to low, variable and localized nutrient availability. Our knowledge of these adaptations is presently quite limited. However, observations of bacteria collected from natural environments and bacteria grown under defined conditions of nutrient limitation have revealed several rather general strategies of response to diminished nutrient supply. These adaptations include: 1) sporulation; 2) formation of miniature cells; 3) derepression of catabolic enzymes; 4) derepression of assimilatory or anabolic enzymes; and 5) changes in uptake and transport activities. These kinds of adaptive response to starvation or nutrient limitation have been well documented (Poindexter 1987; Harder and Dijkhuizen 1983; Matin et al 1989).

There may also be major changes in the motility and chemotactic responsiveness of bacteria during nutrient limited growth. Unfortunately, this possibility seems to have been carefully examined in only one previous study. Terracciano and Canale-Parola (1984) showed that *Spirochaeta aurantia* from carbon-limited chemostat cultures responded to glucose and xylose concentrations 10- to 1,000-fold lower than cells from batch cultures. Maximum enhancement of chemotaxis occurred in cells growing at sugar concentrations in the range of 3 to 4 micromolar. Growth on low levels of glucose stimulated chemotaxis to glucose but not to xylose. These results provide clear evidence for substantial and specific up-regulation of chemotactic responsiveness in *S. aurantia* during growth under carbon limitation.

In parallel chemostat studies, we cultured *R. meliloti* on growth-limiting concentrations of either C or N to determine whether changes in nutrient availability altered the bacterium's commitment to chemotactic exploration. In contrast to *S. aurantia*, 95% to 100% of our *R. meliloti* cells became non-motile within 12-24 h, apparently due to loss of flagella (Robinson et al 1990). Since these non-motile cells had sufficient energy and metabolites to continue dividing at 20-50% of their maximal rate, it seems likely that they had more than enough energy to synthesize and operate their flagella. Tentatively then, this initial loss of motility during growth of *R. meliloti* on limiting C or N seems to be a regulatory response on the part of the bacteria, allocating available nutrients to growth instead of exploration.

In C-limited, but not N-limited, chemostat cultures there was a second major change in the motility of R. meliloti. As steady state growth rates were achieved after about 40-50 h, an increasing percentage of the non-motile cells were replaced by small, highly motile forms of the bacterium, reaching a steady 40% of the population by 96 h (Robinson et al 1990). Very small, highly motile cells of rhizobia, called "swarmers", were first described over a century ago, but never before reproducibly cultured on defined media. Little is known about either the circumstances of swarmer formation or their biological function. Electron micrographs of R. meliloti swarmers show flagellated coccoids 0.1-0.3 μm diameter (Dart and Mercer 1964). Our chemostat cultures contained a mixture of such coccoids and the normal 0.5 x 1.5 μm rods. Bewley and Hutchinson (1920) reported that coccoid swarmers formed when rhizobia were cultured on soil extract agar. After prolonged culture on soil extracts, active swarmer cells became non-motile, a transformation that was reversible upon addition of various carbon sources. Bottomley and Dughri (1989) recently discovered that a substantial proportion (22% to 34%) of the R. leguminosarum cells recoverable from field soils were of the same small dimensions (<0.4 u) as swarmer cells. None of these small cells were culturable or formed nodules. It seems possible that these small cells were the dead remnants of R. leguminosarum swarmers, swarmers that formed during conditions of nutrient-limited growth, but which did not succeed in finding new nutrients before exhausting their internal reserves. We believe that the reversible formation of Rhizobium swarmers in soil extracts, the induction of swarmer formation in C-limited chemostats, and the recovery of large numbers of swarmer-sized cells from native soils all hint that swarmer formation may be an important, chemotaxis-related adaptation of rhizobia to soil environments.

Effects of starvation on motility and chemotaxis

Several studies indicate that transfer to starvation conditions can elicit changes in motility or chemotactic responsiveness. For example, the marine vibrio Ant300 is weakly motile and chemotactic in batch cultures on rich media. But this strain becomes highly motile and chemotactic after transfer to a starvation buffer, with enhanced activity evident 15 h after transfer and persisting for 7 days (Torella and Morita 1982). In contrast, cells of another marine vibrio, strain S14, gradually lost motility during the first 10 h after transfer to starvation conditions, and the motile cells lost responsiveness to various attractants at different rates (Malmcrona-Friberg et al 1990). These observations suggest that the regulation of chemotaxis in response to severe nutrient deprivation is likely to be both diverse and complex, even among isolates inhabiting apparently similar environments.

Cells from early log phase cultures of various R. meliloti strains remain highly motile and chemotactically responsive for about 3 hours after resuspension in starvation buffer (Wei, Robinson and Bauer, unpublished). For strains 2011 and R400, the percentage of motile cells rapidly diminished to zero within 5 h after transfer. For strain RMB7201, however, motility dropped from roughly 80% to 10% over the first 9 h after transfer and then remained at that level for another 18-24 h before fading away. Addition of glucose to the starvation buffer had no effect on these changes in motility, indicating that loss of motility in starvation media is a programmed response rather than a lack of sufficient energy to drive the flagellar motors.

Isolation of mutants with enhanced chemotaxis

Three laboratories, including ours, have recently isolated "hypertaxis" mutants of *R. meliloti*, mutants which form significantly *larger* swarm colonies than the wild-type in semisolid agar plates (Wei, Robinson and Bauer, unpublished; Krupski et al 1985; Ronco 1988; Bauer and Caetano-Anollés, 1990). The formation of larger swarms most likely results from some sort of up-regulation of chemotactic activity. Swarm colony growth is initiated when bacteria placed in the center of a soft agar plate start to consume attractant metabolites. This creates concentration gradients of the attractants that bias the movement of motile cells towards the edge of the plate. Swarm colony growth in semisolid agar is surprisingly complex, with many genetic and environmental factors contributing to migration rate and pattern (Wolfe and Berg, 1989). Thus, there are a variety of ways that mutations might result in the apparent up-regulation or derepression of chemotactic migration.

The NTG-induced hypertaxis mutant isolated by Krupski et al was reported to overproduce flagella and swim about 50% faster than the wild-type (Götz and Schmitt 1987). The Tn5-induced mutant isolated by Ronco (1988) was described only as having greater than normal responsiveness in forming "clouds" on host roots. In our initial studies, we isolated both spontaneous and Tn5-induced hypertaxis mutants of *Bradyrhizobium japonicum* by simple serial enrichment for cells which migrated fastest and ended up at the outer edge of swarm colonies (Bauer and Caetano-Anollés 1990). One of these mutants was found to initiate nodules on soybean about 5 times more efficiently than the wild-type and about 25 times more efficiently than a non-motile mutant isolate. These results encouraged us to undertake a more extensive search for similar mutants in *R. meliloti*. Screening of individual colonies from Tn5 matings has thus far yielded two mutants of *R. meliloti* RMB7201 with the hypertaxis phenotype, while 15 cycles of serial enrichment for spontaneous mutants has yielded three additional hypertaxis mutants. These mutants all generate swarm colonies that increase in size about 1.5 to 3 times faster than the wild-type. Initial observations of our first *R. meliloti* Tn5 hypertaxis mutant indicate that it has rather normal motility and responsiveness: the reason for its enhanced swarm migration is not obvious. The other hypertaxis mutants have just recently been purified and not yet examined in any detail. Clearly, a great deal remains to be done to characterize the nature and molecular origins of hypertaxis behavior in the various mutants. Characterization of such mutants promises to reveal basic control mechanisms that a soil bacterium might use to adjust its motility and chemotactic activity to optimal levels in complex environments.

References

Aguilar, J.M.M., Ashby, A.M., Richards, A.J.M., Loake, G.J., Watson, M.D. & Shaw, C.H. (1988) *J. General Microbiology* **13**, 2741-2746.

Allweis, B., Dostal, J., Carey, K.E., Edwards, T.F. & Freter, R. (1977) *Nature* **266**,448-450.

Ames, P. & Bergman, K. (1981) *J. Bacteriology* **148**, 728-729.

Ames, P., Schluederberg, S.A. & Bergman, K. (1980) *J. Bacteriology* **141**, 722-727.

Armitage, J.P., Gallagher, A. & Johnston, A.W.B. (1988) *Molecular Microbiology* **2**, 743-748.

Bauer, W.D. & Caetano-Anollés, G. (1990) *Plant & Soil* **129**, 45-52.

Bergman, K., Gulash-Hoffee, M., Hovestadt, R.E., Larosiliere, R.C., Ronco , P.G.II & Su, L. (1988) *J. Bacteriology* **170**, 3249-3254.

Bewley, W.F. & Hutchinson, H.B. (1920) *J. Agricultural Science* **10**, 144-165.

Caetano-Anollés, G., Wall, L.G., DeMicheli, A.T., Macchi, E.M., Bauer, W.D. & Favelukes, G. (1988a) *Plant Physiology* **86**, 1228-1235.

Caetano-Anollés, G., Crist-Estes, D.K. & Bauer, W.D. (1988b) *J Bacteriology* **170**, 3164-3169.

Chet, I. & Mitchell, R. (1976) *Annual Review Microbiology* **30**, 221-239.

Dart, P.J. & Mercer, F.V. (1964) *Archives Microbiology* **47**, 344-378.

De Weger, L.A., van der Vlugt, C.I.M., Wijfjes, A.H.M., Bakker, P.A.H.M., Schippers, B. & Lugtenberg, B. (1987) *J. Bacteriology* **169**, 2769-2773.

Dharmatilake, A. (1990) MSc Thesis, Ohio State University, Columbus, OH.

Götz, R., Limmer, N. Ober, J. & Schmitt, R. (1982) *J. General Microbiology* **128**, 789-798.

Götz, R. & Schmitt, R. (1987) *J. Bacteriology* **169**, 3146-3150.

Gulash, M., Ames, P., LaRosiliere, R.C. & Bergman, K. (1984) *Applied & Environmental Microbiology* **48**, 149-152.

Haefele, D.M. & Lindow, S.E. (1987) *Applied & Environmental Microbiology* **53**, 2528-2533.

Howie, W.J., Cook, R.J., & Weller, D.M. (1987) *Phytopathology* **77**, 286-292.

Hunter, W.J. & Fahring, C.J. (1980) *Soil Biology & Biochemistry* **12**, 537-542.

Koshland, D.E. (1981) *Annual Review Biochemistry* **50**, 765-782.

Krupski, G., Gotz, R., Ober, K., Pleier E. & Schmidt, R. (1985) *J. Bacteriology* **162**, 361-366.

Liu, R., Tran, V.M. & Schmidt, E.L. (1989) *Applied & Environmental Microbiology* **55**, 1895-1900.

McNab, R. (1987a) In: *Escherichia coli* and *Salmonella typhimurium*. Cellular and Molecular Biology. FC Neidhardt et al (eds.) American Society Microbiologists, Washington DC. pp 70--83.

McNab, R. (1987b) In: *Escherichia coli* and *Salmonella typhimurium*. Cellular and Molecular Biology. FC Neidhardt et al (eds.) American Society Microbiologists, Washington DC. pp 732-759.

Malek, W. (1989) *Archives Microbiology* **152**, 611-612.

Malmcrona-Friberg, K., Goodman, A. & Kjelleberg, S. (1990) *Applied & Environmental Microbiology* **56**, 3699-3704.

Maxwell, C.A., Hartwig, U.A., Joseph, C.M. & Phillips, D.A. (1989) *Plant Physiology* **91**, 842-847.

Mellor, H.Y., Glenn, A.R., Arwas, R. & Dilworth, M.J. (1987) *Archives Microbiology* **148**, 34-39.

Napoli, C.A. & Albersheim, P. (1980) *J. Bacteriology* **141**, 979-980.

Parke, D., Ornston, L.N., & Nester, E.W. (1987) *J. Bacteriology* **169**, 5336-5338.

Pleier, E. & Schmitt, R. (1989) *J. Bacteriology* **171**, 1467-1475.

Reynolds, P.J., Sharma, P., Jenneman, G.E. & McInerney, M.J. (1989) *Applied & Environmental Microbiology* **55**, 2280-2286.

Robinson, J.B., Bauer, W.D. & Tuovinen, O.H. (1990) Abstract, 1990 annual meeting American Society Microbiologists, Anaheim, CA.

Rolfe, B.G. & Gresshoff, P.M. (1988) *Annual Review Plant Physiology* **39**, 297-319.

Ronco, P.G. II. (1988) MSc Thesis. Northeastern University, Boston MA.

Scher, F.M., Kloepper, J.W., Singleton, C., Zaleska, I, & Laliberte. (1988) *Phytopathology* **78**, 1055-1059.

Shaw, C.H., Ashby, A.M., Brown, A., Royal, C. & Loake, G.J. (1988) *Molecular Microbiology* **2**, 413-418.

Terracciano, J.S. & Canale-Parola, E. (1984) *J. Bacteriology* **159**, 173-178.

Wolfe, A.J. & Berg, H.C. (1989) *Proceedings National Academy Sciences* **86**, 6973-6977.

Ziegler, R.J., Pierce, C & Bergman, K. (1986) *J. Bacteriology* **168**, 785-790.9

Distinct entities between soybean agglutinin receptor and soybean root hair binding site on *Bradyrhizobium japonicum* cell surface

Siu-Cheong Ho

Dept. of Biochemistry, Michigan State University, East Lansing, Michigan 48824

Introduction

Bradyrhizobium japonicum attaches to soybean root in a polar fashion. In previous studies, it was shown that this binding activity was mediated through a carbohydrate-specific mechanism, such that galactose (Gal) inhibited this binding activity (1-3). Lectins from legume hosts were originally proposed to mediate the *Rhizobium* binding process (4-6). Results from recent studies indicate that the soybean agglutinin (SBA) may not play an important role in the attachment of *B. japonicum* to soybean roots. One conspicuous evidence that argues against SBA involvement in *B. japonicum* binding to the soybean roots is that, while Gal inhibits *B. japonicum* binding to soybean roots, N-acetyl-D-galactosamine (GalNAc) does not (1,2). Should SBA be involved in *Rhizobium* attachment, GalNAc should inhibit this interaction. This discrepancy may suggest the involvement of other mechanism that can exquisitely distinguish Gal from GalNAc. The present report is to give further evidence to explain why SBA may not participate in the bacterial attachment process. These results lead to the discovery of another mechanism possibly mediated by a carbohydrate binding protein present in *B. japonicum* recognizing the glycoconjugate on the legume host cell surface.

SBA receptor and soybean root binding site

The localization of SBA receptor in *B. japonicum* cells has been examined by FITC-labeled SBA (7,8). Fluorescent labeling was observed at one pole of the rod-shaped bacteria indicating an uneven distribution of SBA receptor on the bacterial surface. This receptor has been characterized to be the capsular polysaccharides (CPS) of the bacteria (9,10). These results are in agreement with the ultrastructural studies by electron microscopy (11-13). SBA coupled with colloidal gold particles was labeled on the bacterial surface polysaccharide at one pole of the cell. Transmission electron microscopic studies of the cross-section of *B. japonicum* indicated that the two poles of the bacterium are clearly distinct from each other. The nucleoid portion which mainly consists of the bacterial chromosome and other cytoplasmic constituents, is located on one end. The other end mainly contains the granules which are identified to be glycogen and poly-ß-hydroxybutyrate granules. The CPS was localized at the surface of the nucleoid portion. When the bacteria were labeled with FITC-SBA and then allowed to interact with the cultured soybean SB-1 cells, the bacteria attached to the soybean cells at one pole (Figure 1). The pole that carried the FITC-SBA was localized at the other pole of the bacteria different from the pole mediating attachment. Similar results were obtained when the bacteria were allowed to attach to the SB-1 cells prior to FITC-SBA labeling. FITC-SBA showed negligible binding to the cultured soybean SB-1 cell surface. This result indicates that the cultured soybean cell surface

Figure 1. FITC-SBA labeling of B. japonicum *after attachment to cultured soybean SB-1 cell surface.*

did not seem to be able to recognize internal Gal residue. Upon partial digestion of the SB-1 cell wall with pectinase, FITC-SBA labeled on SB-1 cell surface, presumably exposing the Gal residues by pectinase. From these results, it is suggested that the bacteria bind to cultured soybean SB-1 cells with one pole of the bacterium. This pole is distinctly different from the pole that contains SBA receptor. Therefore, the SBA receptor does not seem to participate in the binding process. Pueppke examined the possibility of SBA involved in *B. japonicum* binding to the soybean roots (14). It was shown that the addition of SBA to completely saturate the SBA receptor on the bacterial surface failed to have substantial inhibition of the initial adsorption of the bacteria to the soybean root surface. The results obtained in this laboratory confirmed this observation. It is therefore speculated that the complementary ligand-receptor mediating *B. japonicum*-soybean interaction is different from the SBA-CPS complementary pair.

When *B. japonicum* cells were cultured in semi-solid agar medium, they interacted to each other through one pole of the cell, forming rosette structures of 10-20 cells (Figure 2). This phenomenon is called star formation. This star formation activity has been observed in bacterial strains, like *Rhizobium lupini*, and *Pseudomonas echinoides* (15,16). In *B. japonicum*, this star-like structure is quite stable and can withstand vigorous mechanical shearing forces. When the star-forming bacteria were cultured in liquid culture for 24 h and then labeled with FITC-SBA, the fluorescence appeared at the pole of the cell away from the point of cell-cell contact (Figure 3). The center of aggregation was free of FITC-SBA labeling. Ultrastructural studies of the star-forming bacteria clearly indicated that the granules of the bacteria were oriented close to the center of cell-cell contact, while the nucleoid portion of the bacteria is free away from the attachment site (12). From these data, it can be concluded that SBA receptor is located at a bacterial pole that is different from the pole that participates in star formation.

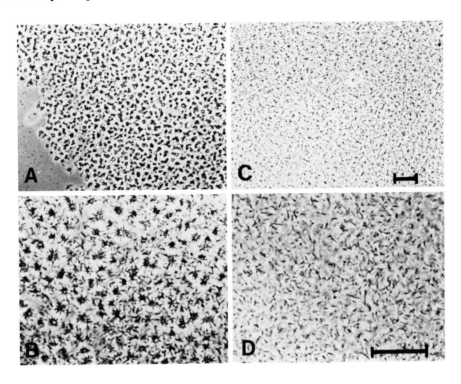

Figure 2. Lac inhibition of star formation of B. japonicum culture in YEMG agar medium. (A,B) control without Lac; (C,D) in the presence of 50 mM Lac. Bar represents 100 μm.

These results are in accordance with that of the heterotypic interaction between *B. japonicum* and culture soybean SB-1 cells. FITC-SBA labeled the bacterial end at the nucleoid portion. However, the poles that both involve in homotypic star formation and heterotypic *B. japonicum*-soybean interactions are identified to be the same at the granular portion of the bacteria. This view point is further supported by the ultrastructural identification of the polar bodies at this end of the bacteria (12,13). Electron microscopy revealed that these structures contain fibrillar structures, similar to fimbriae. Vesper et al. (17) purified pili from *B. japonicum* and demonstrated two polypeptides of molecular weights 21,000 and 18,000. Antiserum generated against this pilus fraction labeled *B. japonicum* at one pole and blocked bacteria binding to soybean roots and nodulation. These data suggest that pili may be involved in the *B. japonicum* binding to the soybean roots. Consistent with this notion, mutants of *B. japonicum* defective in attachment were shown to have a concomitant reduction in pilus formation. However, the sugar specific recognition properties of the isolated pili has not been demonstrated (17).

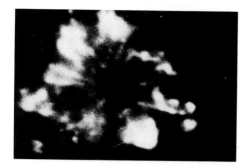

Figure 3. FITC-SBA labeling of a star-like cluster of B. japonicum *cells. The center of cell-cell contact is devoid of labeling.*

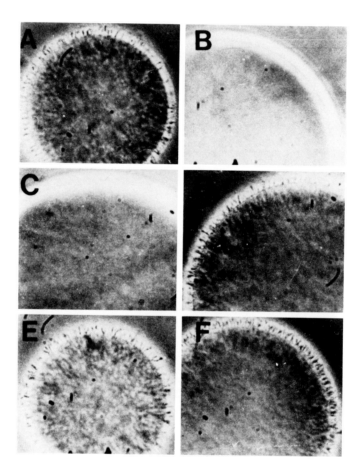

Figure 4. Saccharide-specific inhibition of B. japonicum *to Lac-beads. (A) control without saccharide; others with (B) Lac; (C) Gal; (D) GalNAc; (E) Glc and (F) Man. The final concentration of all saccharides was 50 mM.*

Saccharide-specific binding of *B. japonicum*

To examine the possibility that *B. japonicum* possesses ability for carbohydrate-specific recognition, *B. japonicum* culture was tested to bind to synthetic beads derivatized with various saccharides (18). Results indicate that the bacteria could bind to acrylamide beads that were covalently coupled with lactose (4-ß-D-galactopyranosyl-D-glucose), but not to beads coupled with maltose (4-O-α-D-glucopyranosyl-D-glucose) or sucrose (α-D-glucopyranosyl-ß-D-fructofuranoside) (19).

Similar to the bacterial binding to the soybean roots and cultured soybean SB-1 cells, the bacteria also attached to the lactose (Lac) beads in a polar fashion (Figure 4). This attachment could be inhibited by incubating the bacteria with the beads in the presence of 50 mM Lac or Gal. Glucose (Glc), xylose (Xyl) or mannose (Man) did not show any inhibitory effect. GalNAc, a derivative of Gal at the C-2 position, at a concentration of 50 mM also showed no inhibition. These results indicate that a carbohydrate-recognition mechanism is present in *B. japonicum* that can exquisitely distinguish Gal from other saccharides.

In consistent with this notion, the star-forming activity of *B. japonicum* is mediated by the same saccharide-specific mechanism. Since it could be inhibited by culturing the bacteria in the presence of Lac or Gal (Figure 2). GalNAc did not show any effect. By comparing the saccharide specificity in four assay systems, it is clear that the saccharide specificity is maintained throughout all assay systems. Therefore, it is likely that all of these binding activities are mediated through the same mechanism involving a carbohydrate-specific recognition machinery located on the bacterial surface. A carbohydrate-binding protein was then proposed to be present on *B. japonicum* cell surface to mediate all four binding activities of *B. japonicum* (Figure 5).

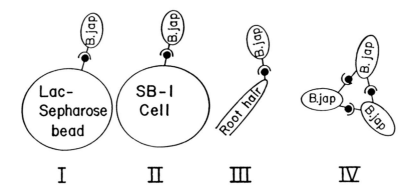

Figure 5. Models illustrating the role of BJ38 in the binding of B. japonicum *to: (I) Lac-Sepharose beads; (II) cultured soybean SB-1 cells; (III) soybean root hair; (IV) other* B. japonicum *cells leading to star formation. BJ38; carbohydrate ligand.*

Isolation of a carbohydrate-binding protein

B. japonicum cells were cultured in YEMG medium in suspension culture to late logarithmic phase at which *B. japonicum* expressed maximum binding activity. Cells were harvested and ruptured by a French press. The cell fraction was extracted and subjected to affinity chromatography on a column of Sepharose coupled with Lac (20). After extensive washing to remove the unbound materials with PBS, the bound materials were eluted with 0.1 M Lac. Multiple bands were identified by SDS-PAGE analysis in the purified sample (Figure 6). When this sample was rechromatographed on the affinity column, upon SDS-PAGE analysis, one predominant polypeptide was observed by silver stain indicating apparent homogeneity of the sample. When this protein was radioiodinated by chloramine T method, and reisolated from the affinity column, SDS-PAGE and autoradiography showed one single band with mobility corresponding to that of the silver-stained component. This result confirmed the Lac specificity of the protein, which is indicated by specific binding to the affinity column containing Lac and by specific elution by Lac. By comparing the mobilities of the molecular weight markers upon SDS-PAGE analysis, this protein corresponded to a molecular weight of 38,000. Therefore, it is designated as BJ38.

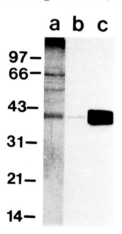

Figure 6. SDS-PAGE analysis of samples isolated from B. japonicum *cell extract after purification from a Lac-Sepharose column. Samples after one cycle (a) and second cycle (b) of affinity chromatography; (c) autoradiography of the radioiodinated sample from (b).*

The saccharide specificity of this protein was further analyzed by stepwise elution of this protein from the affinity column of Lac-Sepharose. After radioiodination, the ^{125}I-BJ38 was bound to the affinity column, the column was developed stepwise with Glc, Man, GalNAc and Gal at a concentration of 50 mM (Figure 7). ^{125}I-BJ38 could not be eluted from the affinity column by Glc, Man, or even GalNAc. However, when 50 mM Gal was applied onto the column, ^{125}I-BJ38 was eluted out. These results clearly demonstrated that ^{125}I-BJ38 was able to differentiate Gal from other saccharides. It could exquisitely distinguish between Gal and GalNAc. More importantly, the saccharide specificity of ^{125}I-BJ38 is in agreement with that of the *B. japonicum* binding to the soybean roots. It is, therefore, suggested that this carbohydrate binding protein may mediate the saccharide-specific binding between *B. japonicum* and soybean roots.

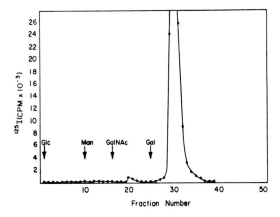

Figure 7. Saccharide-specificity of ^{125}I-BJ38 as determined by stepwise elution from a Lac-Sepharose affinity column. After ^{125}I-BJ38 was bound on the affinity column, the column was developed by sequential elution with Glc, Man, GalNAc and Gal at 50 mM.

After *B. japonicum* wild type strain was mutagenized with N-methyl-N'-nitro-N-nitrosoguanidine, two mutants, N4 and N6, were isolated based on their deficiency in binding to cultured soybean SB-1 cells (19,20). These mutants showed greatly diminished in binding activity under all four binding assays to (a) soybean roots, (b) cultured soybean SB-1 cells, (c) Lac-beads and (d) other *B. japonicum* (autoagglutination or star formation). When these mutants were extracted for BJ38 isolation, after Lac-Sepharose affinity chromatography, no BJ38 could be detected from either mutants. These results concluded that mutants that were defective in expressing BJ38 were defective in all binding activities of *B. japonicum*. This conclusion indirectly supports the hypothesis that BJ38 is important in mediating all four binding properties of *B. japonicum*.

Conclusion

B. japonicum attaches to soybean root through a carbohydrate-specific mechanism. SBA does not seem to play an indispensable role in this attachment process. This conclusion is based on three critical observations. First, Gal inhibited *B. japonicum* binding to soybean roots, but GalNAc did not. Both saccharides are potent hapten inhibitors for SBA. Second, saturable amounts of SBA to the lectin receptor sites on *B. japonicum* cell surface did not block *B. japonicum* binding to soybean roots. Third, FITC-SBA bound to one pole of *B. japonicum* which did not involved in soybean root binding.

Further analysis indicates that *B. japonicum* cells possess carbohydrate recognition activities. They can bind to synthetic beads coupled with Lac and can interact with each others through a saccharide specific manner. The saccharide specificity of these binding activities is similar to that of the bacterial binding to soybean roots and cultured soybean SB-1 cells. From these results, it is proposed that *B. japonicum* cells exhibit carbohydrate recognition. This proposal leads to the identification of a carbohydrate binding protein, namely BJ38. This protein shares the same saccharide specificity as that of *B. japonicum* binding activities. It is, therefore, hypothesized that BJ38 mediates all the saccharide specific binding activities of *B. japonicum*. This hypothesis is further supported by the fact that two binding-deficient mutants are also deficient in expressing functional BJ38.

References

1.   Ho, S.-C., Ye, W., Schindler, M., Wang, J.L. (1988) J. Bacteriol. **170**, 3882-3890.

2.   Vesper, S.J., Bauer, W.D. (1985) Symbiosis **1**, 139-162.

3.   Halverson, L.J., Stacey, G. (1985) Plant Physiol. **77**, 621-625.

4.   Hamblin, J., Kent, S.P. (1973) Nature New Biol. **254**, 28-30.

5.   Dazzo, F.B. (1981) J. Supramol. Structure Cell Biochem. **16**, 29-41.

6.   Kato, B., Maruyama, Y., Nakamura, M. (1981) Plant and Cell **22**, 759-771.

7.   Bohlool, B.B., Schmidt, E.L. (1974) Science **185**, 269-271.

8.   Bohlool, B.B., Schmidt, E.L. (1976) J. Bacteriol. **125**, 1188-1194.

9.   Mort, A.J., Bauer, W.D. (1980) Plant Physiol. **66**, 163-168.

10.   Bhuvaneswari, T.V., Pueppke, S.G., Bauer, W.D. (1977) Plant Physiol. **60**, 486-491.

11.   Bal, A.K., Shantharam, S., Ratnam, S. (1978) J. Bacteriol. **133**, 1393-1400.

12.   Tsien, H.C., Schmidt, E.L. (1977) Can. J. Microbiol. **23**, 1274-1284.

13.   Tsien, H.C. (1982) In: Nitrogen Fixation, Broughton, W.J. (ed.), Clarendon Press, Oxford, Vol. 2, pp. 182-198.

14.   Pueppke, S.G. (1984) Plant Physiol. **75**, 924-928.

15.   Heumann, W. (1968) Mol. Gen. Genet. **102**, 132-144.

16.   Heumann, W. (1956) Arch. Mikrobiol. **24**, 362-395.

17.   Vesper, S.J., Bauer, W.D. (1986) Appl. Environ. Microbiol. **52**, 134-141.

18.   Baues, R.J., Gray, G.R. (1977) J. Biol. Chem. **252**, 57-60.

19.   Ho, S.-C., Wang, J.L., Schindler, M. (1990) J. Cell Biol. **111**, 1631-1638.

20.   Ho, S.-C., Schindler, M., Wang, J.L. (1990) J. Cell Biol. **111**. 1639-1643.

Rhizobium lipopolysaccharides; their structures and evidence for their importance in the nitrogen-fixing symbiotic infection of their host legumes

Russell W. Carlson, U. Ramadas Bhat and Brad Reuhs

Complex Carbohydrate Research Center, Univ. of Georgia, Athens, GA 30602.

Introduction

The purpose of this brief chapter is to describe what is known about the structure of *Rhizobium* lipopolysaccharides (LPSs), and to summarize the evidence showing that these molecules are important in establishing a nitrogen-fixing symbiosis between a *Rhizobium* and its legume host. Work regarding other *Rhizobium* surface polysaccharides, *e.g.* extracellular polysaccharides (EPSs), is discussed when it relates to, or clarifies, the role of LPS in symbiosis. The readers are referred to recent and comprehensive reviews describing *Rhizobium* polysaccharies, their genetics and possible roles in symbiosis (Gray & Rolfe, 1990; Noel, 1991).

The Structures of *Rhizobium* LPSs

General structural features

Rhizobium LPSs, as with the more extensively studied LPSs from enteric bacteria, consist of a polysaccharide and a lipid (lipid A). The polysaccharide fraction is comprised of a larger molecular weight molecule known as the O-antigenic polysaccharide or O-chain, and a core oligosaccharide. The lipid A usually consists of an oligosaccharide (usually a disaccharide) to which are attached fatty acyl residues. The polysaccharide portion is attached to the lipid A via an eight carbon sugar known as 3-deoxy-D-*manno*-2-octulosonic acid (Kdo). A typical LPS preparation from *Rhizobium* or other gram-negative bacteria consists of a mixture of complete and incomplete LPSs. Complete forms of the LPS contain lipid A, core oligosaccharide and O-antigen polysaccharide. Incomplete forms of the LPS contain lipid A and core oligosaccharide but lack the O-antigen polysaccharide. Other than these general characteristics, *Rhizobium* LPSs have quite different structural characteristics from those of enteric bacteria.

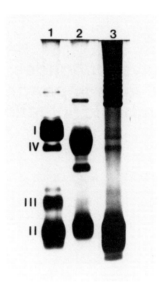

Figure 1. *A DOC-PAGE silver-stained gel of three different* Rhizobium *LPSs. Well 1 contains LPS from* R. leguminsoarum bv. phaseoli *CE3, well 2 from* R. leguminosarum bv. trifolii *ANU843 and well 3 from* R. leguminosarum bv. viciae 128C53. *The banding pattern can vary greatly among various* Rhizobium *strains. The ladder pattern for the LPS from bv. viciae (well 3) is characterisitic of LPSs which have an O-antigen polysaccharide that varies greating in the number of repeating oligosaccharide unit. Each band is separated from the one below it in that it contains an O-antigen polysaccharide with one additional repeating unit.*

Rhizobium leguminosarum LPSs

The LPSs are prepared by hot phenol-water extraction and further purified by gel-filtration chromatography (Carlson *et al.* 1978; Westphal & Jann, 1965). The heterogeneity inherent in LPS preparations, from *Rhizobium* and other gram-negative bacteria, is revealed by polyacrylamide gel electrophoresis in the presence of sodium dodecyl sulfate (SDS-PAGE) or deoxycholate (DOC) (DOC-PAGE) followed by silver-staining (Carlson, 1984; Hitchcock & Brown, 1983; Krauss *et al.* 1988). A DOC-PAGE of *Rhizobium* LPSs is shown in Figure 1. The higher molecular weight forms of the LPS, LPSs I and IV, contain the O-antigenic polysaccharide, core and lipid A; while the lower molecular weight LPSs, LPS II and III, contain only the core oligosaccharide and lipid A. These results were obtained by immunoblots using LPS antisera and by chemical analysis of the separated LPS forms (Carlson *et al.* 1987b; Carlson *et al.* 1987a; Carlson & Lee, 1983). The LPS I and IV were separated from LPS II and III by gel filtration chromatography in the presence of DOC. This procedure was developed by (Peterson & McGroarty, 1985) for LPSs from enteric bacteria.

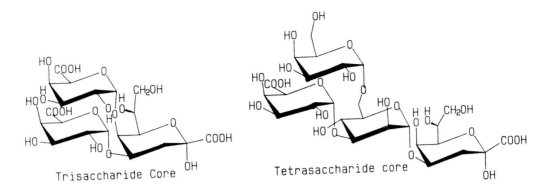

Figure 2. Structures of the LPS core oligosaccharides from strains of R. leguminosarum.

Individual LPS components were isolated by mild acid hydrolysis, which breaks the Kdo bond linking the polysaccharide to the lipid A, followed by gel-filtration chromatography of the water-soluble polysaccharide (Carlson et al. 1988; Carlson, 1984; Carlson et al. 1990; Ryan & Conrad, 1974). The lipid A released by mild acid hydrolysis precipitated and was purified by centrifugation and/or extraction with chloroform. The O-antigenic polysaccharide and two core oligosaccharides were purified by gel-filtration (Bio-Gel P2 in 1% acetic acid) (Carlson, 1984; Carlson et al. 1990). The structures of the individual components were then investigated using GC, GC-MS, FAB-MS and NMR techniques.

The O-antigenic polysaccharides are complex and highly variable among strains of *R. leguminosarum.* Sugar components can consist of numerous methylated glycosyl residues, methylated glycosamines, 6-deoxyglycosamines, heptose, uronic acid, and Kdo (Carlson, 1984; Carlson *et al.* 1978; Carlson *et al.* 1987; Carlson *et al.* 1987b). The Kdo residue is found at the reducing end of the O-antigenic polysaccharide from all strains of *R. leguminosarum* examined to date. Strains of *R. leguminosarum* which nodulate the same host legume can have very different O-antigenic polysaccharides (Carlson, 1984; Carlson *et al.* 1978).

Mild acid hydrolysis of *R. leguminosarum* LPSs releases two oligosaccharides, a tri- and a tetrasaccharide. The structures of these two oligosaccharides have been determined for both *R. leguminosarum* bv. trifolii (Carlson *et al.* 1988; Hollingsworth *et al.* 1989c; Hollingsworth *et al.* 1989a; Bhat *et al.* 1991b), bv. phaseoli (Carlson *et al.* 1990; Bhat *et al.* 1991b) and bv. viciae (Carlson, unpublished). These structures are shown in Figure 2 and are identical for all *R. leguminosarum* strains examined. Occasionally the tetrasaccharide component may be acetylated as reported for the bv. trifolii tetrasaccharide (Hollingsworth *et al.* 1989c). It should also be noted that mild acid hydrolysis releases monomeric Kdo. Thus, a complete LPS molecule from these strains contains at least four Kdo residues; one each at the reducing ends of the O-antigen polysaccharide, the core tri- and tetrasaccharides, and monomeric Kdo. It is not yet known how these various LPS components are arranged in the intact molecule.

The structure of the lipid A portion of these LPSs is under investigation. A unique feature of these LPSs, compared with the LPSs enteric bacteria, is the presence of a very long chain fatty acyl residue, 27-hydroxyoctacosanoic acid (Hollingsworth & Carlson, 1989b). The other major fatty acyl residue is 3-hydroxymyristic acid (Carlson *et al.* 1987b; Bhat *et al.* 1991a) which is common to many LPSs. Other fatty acyl residues include 3-hydroxypalmitic, 3-hydroxystearic and 3-hydroxypentacosanoic acids (Carlson *et al.* 1987b; Bhat *et al.* 1991a). These fatty acyl residues are in amide or ester linkages to the sugar backbone thereby forming the lipid A portion of these LPSs. One report indicates that a fraction of the lipid A from bv. trifolii consists of 27-hydroxyoctacosanoic acid N-linked to glucosaminuronic acid (Hollingsworth & Lill-Elghanian, 1989d). Work in our laboratory has not been able to repeat the results of this report, however it is possible that this lipid A fraction is present as a very minor component which we did not detect. The major glycosyl components of the LPS lipid A fractions from all strains of *R. leguminosarum* that have been examined in our laboratory are galacturonic acid and glucosamine. A preliminary report indicates that the galacturonic acid is a-linked to C4 of the glucosamine, and that 27-hydroxyoctacosanoic acid is N-linked to glucosamine while the other fatty acyl residues are ester linked to both galacturonic acid and glucosamine (Bhat & Carlson, 1990). There is also a significant level of heterogeneity in the fatty acylation pattern of this lipid A. Phosphate is not detected in the lipid A. Further work is in progress to determine the structure of this lipid A.

Bradyrhizobium japonicum *LPSs*

The *B. japonicum* strains are divided into several serogroups. Generally the LPSs from strains of all serogroups are extracted into the phenol layer during phenol-water extraction (Carrion *et al.* 1990). An exception is the LPS from strain USDA110 which is extracted into both the phenol and water layers (Carrion *et al.* 1990). Further purification is accomplished using dialysis, protease and nuclease digestion, and gel filtration chromatography (Carrion *et al.* 1990). Extraction of LPSs into the phenol phase during phenol-water extraction, while unusual, has also been reported for the LPSs from *Rhodopseudomonas palustris* (Whittenburry & McLee, 1967) and *Brucella abortus* (Kreutzer *et al.* 1979; Marx *et al.* 1983).

Mild acid hydrolysis breaks the Kdo glycosidic bond which joins the polysaccharide to the lipid A portion of *B. japonicum* LPSs. However, the hydrolysis conditions require a minimum of 5 hours at 100 °C rather than the usual 1 hour time period for other *Rhizobium* LPSs (Carrion *et al.* 1990; Puvanesarajah *et al.* 1987). The lipid A precipitates and was purified by centrifugation. The polysaccharide portion of the LPS was further purified by gel filtration chromatography. Enrichment for the core oligosaccharide was accomplished by using LPS from mutants which lack the O-antigen polysaccharide (Puvanesarajah *et al.* 1987) and by using isolated LPS II which does not contain the O-antigen polysaccharide (Carrion *et al.* 1990).

The O-antigen polysaccharides from the few *B. japonicum* strains which have been examined vary in their glycosyl components (Carrion *et al.* 1990; Puvanesarajah *et al.* 1987). Typical sugars include fucose, fucosamine, glucose, and quinovosamine (Carrion *et al.* 1990; Puvanesarajah *et al.* 1987). NMR analysis of the O-antigen polysaccharide from strain 61A123 reveals that it is heavily N- and O-acetylated (Carrion *et al.* 1990) which may explain why these LPSs are extracted into the phenol phase during phenol-water extraction. The core region contains 4-O-methylmannose, mannose, glucose and Kdo (Carrion *et al.* 1990). Present results indicate that these components are common to the core region of LPSs from strains representing three distinct serogroups of *B. japonicum* (Carrion *et al.* 1990; Puvanesarajah *et al.* 1987; Carlson *et al.*, unpublished). The lipid A protion of all these LPSs contain 2,3-diaminodideoxyglucose (Carrion *et al.* 1990; Carlson *et al.*, unpublished). Thus, it is likely that the LPSs from the different *B. japonicum* serogroups have a common lipid A and core oligosaccharide region, but vary in their O-antigen polysaccharides. Further structural investigations on *B. japonicum* LPSs are in progress.

Rhizobium meliloti *LPSs*

These LPSs were purified as described above for *R. leguminosarum* LPSs. Analysis by PAGE indicates that these LPSs naturally lack a long O-antigen polysaccharide and consist primarily of the lipid A and core oligosaccharide (Urbanik-Sypniewska *et al.* 1989). Composition results indicate that *R. meliloti* LPSs contain relatively large amounts of Kdo (Carlson, 1982; Urbanik-Sypniewska *et al.* 1989; Zevenhuizen *et al.* 1980). This result is consistent for an LPS that lacks a long O-antigen polysaccharide. Other major glycosyl components are uronic acid, glucose, and glucosamine (Carlson, 1982; Urbanik-Sypniewska *et al.* 1989; Zevenhuizen *et al.* 1980). In one case, phosphate and small amounts of various methylated 6-deoxyhexoses have been reported (Urbanik-Sypniewska *et al.* 1989) Glucosamine is the glycosyl component of the lipid A (Urbanik-Sypniewska *et al.* 1989), and 3-OH-14:0 and 27-OH-28:0 are the major fatty acyl components (Bhat *et al.*, 1991a).

Rhizobium fredii *LPSs*

R. fredii are fast-growing symbionts of soybean. The one published report on *R. fredii* LPS indicates that, as with *R. meliloti*, Kdo is a major glycosyl component, together with uronic acid, glucose and galactose (Carlson & Yadav, 1985). Also, glucosamine is the only glycosyl residue of the lipid A portion of these LPSs (Bhat *et al.* 1991a). Preliminary investigations in our laboratory indicate that the core regions of *R. fredii* and *R. meliloti* LPSs are very similar in structure (Carlson and Reuhs, unpublished). Recently a polysaccharide containing Kdo and galactose has been purified from the LPS preparations of *R. fredii* strains (Reuhs and Carlson, unpublished). The structure of this polysaccharide is being determined in our laboratory. There is also evidence that a similar polysaccharide is present in the LPS preparations of *R. meliloti* strains (Reuhs and Carlson, unpublished).

The Relationship of Rhizobiaceae *LPSs to one another and to the LPSs from other gram-negative bacteria*

Based on structural and compositions studies of *Rhizobium* core oligosaccharides and lipid A's, *Rhizobiaceae* LPSs can be divided into three types (Bhat *et al.* 1991a).

These types are shown in Table I. Type I includes all strains from *R. leguminosarum* biovars viciae, phaseoli and trifolii. Type II includes all *B. japonicum* strains, and type III consists of *R. meliloti* and *R. fredii* strains. In the case of *R. meliloti* and *R. fredii*, there are differences in the fatty acid composition of the lipid A's (Bhat *et al.* 1991a). That *R. fredii* strains are more closely related to *R. meliloti* than other *Rhizobium* species is also supported by DNA homology studies (Gamo *et al.* 1990).

Table I. The Relationship of Rhizobiaceae *Lipopolysaccharides.*

Type	*Rhizobium* Species	Lipid A Components			Core Components					
		GlcN	GalA	DAG	Kdo	Gal	Man	GalA	Glc	4MeMan
I	R. leg. bv. viciae	+	+		+	+	+	+		
	R. leg. bv. trifolii	+	+		+	+	+	+		
	R. leg. bv. phaseoli	+	+		+	+	+	+		
II	B. japonicum 61A123		+	+		+		+	+	
	B. japonicum USDA110		+	+		+		+	+	
	B. japonicum 61A101c		+	+		+		+	+	
III	R. meliloti	+			+	Not determined as yet.				
	R. fredii	+			+	Not determined as yet.				

The discovery of the very long chain fatty acid, 27-hydroxyoctacosanoate, in *Rhizobium* LPSs (Hollingsworth & Carlson, 1989b) has led to a reexamination of many LPSs for the presence of this fatty acid (Bhat *et al.* 1991c; Bhat *et al.* 1991a). All members of *Rhizobiaceae* family contain this fatty acyl residue in their LPSs, except *Azorhizobium caulinodans* (Bhat *et al.* 1991a). *Rhizobiaceae* species examined thus far include *R. leguminosarum, R. meliloti, R. loti, R. fredii, R. galegae, B. japonicum, Bradyrhizobium* sp. (*Lupinus*), *Agrobacterium tumefaciens, A. radiobacter, A. rhizogenes* and *A. rubi* (Kneen *et al.* 1990). Recent DNA-rRNA hybridization studies show that *Azorhizobium caulinodans* is not closely related to other members of the *Rhizobiaceae* but is closely related to *Xanthobacter autotrophicus* and *Xanthobacter flavus* (Dreyfus *et al.* 1988). Thus it is possible that *Azorhizobium caulinodans* should not be classified as being in the *Rhizobiaceae* family. Other bacteria whose LPSs contain 27-hydroxyoctacosanoate belong to the alpha-2 subgroup of *Proteobacteria* (Bhat *et al.* 1991c). The alpha-2 subgroup of *Proteobacteria* which have this fatty acid residue in their LPSs include all members of the *Rhizobiaceae* family as well as *Rhodopseudomonas viridis, Rhodopseudomonas palustris, Nitrobacter winogradsky, Nitrobacter hamburgensis, Pseudomonas carboxydovorans, Brucella abortus,* and the cat scratch disease bacterium (Bhat *et al.* 1991c). There are members of the alpha-2 subgroup which do not contain 27-hydroxyoctacosanoate, however all bacteria which do contain this fatty acyl residue belong to this subgroup of *Proteobacteria* (Bhat *et al.* 1991c).

The function of this very long chain fatty acyl residue is not known. Its chain length gives it the possibility of extending through the entire outer membrane. The location of the hydroxyl group at the penultimate carbon may be essential for interacting with components of the inner leaflet of the outer membrane, such as phospholipids and proteins. This interaction may contribute to increasing the stability and rigidity of the outer membrane of these bacteria. Perhaps this increased outer membrane stability is essential for the survival of these bacteria in a natural, but hostile, environment; *e.g.* remaining viable while within their host's cell.

Evidence for the Importance of Lipopolysaccharides in *Rhizobium*-Legume Symbioses

There are a number of early reports which implicate that *Rhizobium* LPSs play an important role in symbiotic infection (for a review see Carlson, 1982). These reports were directed toward the possibility that LPSs were involved in the specific attachment of the symbiont to the host plant. This early work will not be discussed in detail. However, it should be noted that recent reports on the *R. meliloti*-alfalfa suggest that LPSs are involved in the specific attachment of this symbiont to its host (Caetano-Anolles & Favelukes, 1986; Lagares & Favelukes, 1988).

Rhizobium *LPS Mutants*

The first *Rhizobium* symbiotic mutants which were defective in their LPSs were reported by Noel and coworkers (Noel *et al.* 1986). These mutants are symbiotically defective in the development of infection threads, *i.e.* they abort (Noel *et al.* 1986). These *R. leguminosarum* bv. phaseoli (symbiont of bean) mutants were found, by PAGE analysis, to lack the higher molecular weight forms of their LPS (LPS I and IV, see Figure 1) indicating that they did not possess their O-antigenic polysaccharide (Noel *et al.* 1986). Subsequently this was confirmed by chemical analysis of their isolated LPSs (Carlson *et al.* 1987b). Two of these mutants were found to be defective in the structures of their core oligosaccharides in that the tetrasaccharide core was missing the galacturonsyl residue in one case, and both the galacturonosyl and galactosyl residues in the second case (Carlson *et al.* 1990; Bhat *et al.* 1991b). These structures are shown in Figure 3. Presumably the defect in the core region prevents the addition of the O-antigen polysaccharide. These mutants produce EPSs that are identical to the parent EPS.

Mutants with defective lipopolysaccharides have also been reported for *R. leguminosarum* bv. trifolii (Brink *et al.* 1990) and bv. viciae (Priefer, 1989), and for *B. japonicum* (Puvanesarajah *et al.* 1987; Stacey *et al.* 1991). In all these mutants, the LPSs lack the O-antigenic polysaccharide.

The phenotype of each mutant varies depending on whether the host forms a determinate or indeterminate nodule. In the case of *R. leguminosarum* bv. trifolii and bv. viciae, which form indeterminate nodules on clover and pea respectively, the mutants form normal infection threads, but are defective in the release of the bacteria into the root cortex cells (Brink *et al.* 1990; Priefer, 1989). In the case of the *B. japonicum* and *R. leguminosarum* bv. phaseoli, which form determinate nodules on soybean and bean respectively, the mutants fail in the formation of proper infection threads (Noel *et al.* 1986; Stacey *et al.* 1991). It is also known that EPSs are important in the formation of fully effective indeterminate nodules, but are not necessary for the formation of effective indeterminate nodules (Borthakur *et al.*

1986; Law *et al.* 1982; Diebold & Noel, 1989). In fact, for *R. leguminosarum* bv.

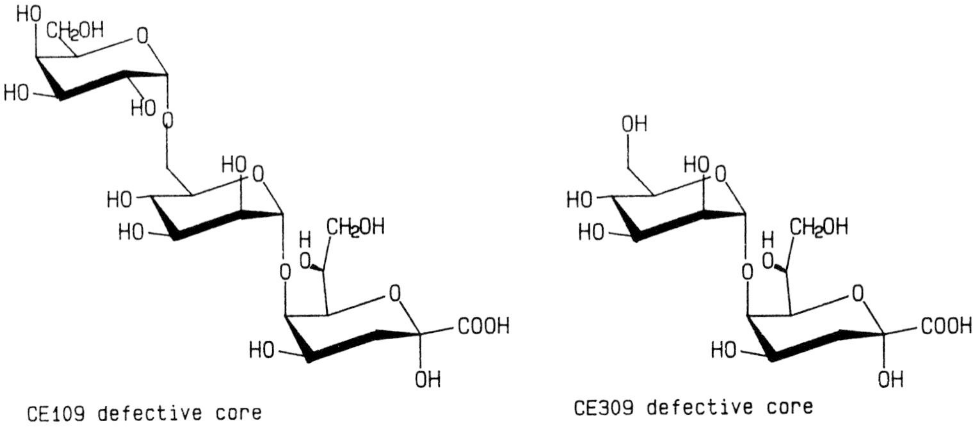

CE109 defective core   CE309 defective core

Figure 3. Structures of the defective core oligosaccharides from R. leguminosarum bv. phaseoli mutants that have LPSs which lack their O-antigen polysaccharides and are symbiotically defective.

trifolii, *R. meliloti* and NGR234 (with an indeterminate host), purified EPS from the parent symbiont, when added to the host-EPS⁻ mutant combination, can restore normal nodulation (Djordjevic *et al.* 1987; Battisti *et al.* 1990). Thus the role of a particular *Rhizobium* polysaccharide is dependent on whether the symbiotic system forms determinate or indeterminate nodules. Lipopolysaccharides are required for both types while EPSs are required only in the latter case. The importance of EPSs for indeterminate but not determinate nodule formation is supported by recent reports showing that EPS⁻ mutants of the broad host range *Rhizobium*, NGR234, form nitrogen-fixing nodules on determinate hosts but are Nod⁺Fix⁻ on indeterminate hosts (Chen *et al.* 1985). The same results have been obtained for *Rhizobium loti* EPS⁻ mutants in determinate and indeterminate hosts (Hotter & Scott, 1991).

Several recent reports have shown that some *R. meliloti* mutants which are defective in EPS synthesis (both EPS I and EPS II) are still able to form nitrogen-fixing nodules (Williams *et al.* 1990a; Williams *et al.* 1990b; Putnoky *et al.* 1988; Putnoky *et al.* 1990). Thus, this appears to be a case in which a symbiont that forms an indeterminate nodule does not require EPS. Mutants of these EPS⁻ strains that form ineffective nodules are reported to have defects in their LPSs (Williams *et al.* 1990a; Williams *et al.* 1990b; Putnoky *et al.* 1990). It is concluded from these reports that *lps* genes are affected in these EPS⁻ mutants and that LPS from the parent EPS⁻

strain can functionally substitute for the missing EPS. However, these conclusions may be premature since the LPSs from these EPS⁻ strains have not been adequately characterized.

While the presence of the O-antigenic polysaccharide is important for symbiosis, there does not appear to be a great deal of stringency on the structural requirements of this polysaccharide. *Rhizobium* strains which have the same host can have widely varying O-antigen polysaccharide structures (Carlson *et al.* 1978; Carlson, 1982). In addition an *R. leguminosarum* bv. trifolii mutant which lacks its O-antigen polysaccharide can be complemented using DNA from an *R. leguminosarum* bv. phaseoli strain which encodes genes for the synthesis of the bv. phaseoli O-antigen polysaccharide. The resulting transconjugate forms normal nitrogen-fixing nodules on clover (the normal bv. trifolii host) even though it contains an LPS that has bv. phaseoli O-antigen sugars and does not have normal bv. trifolii O-antigen polysaccharide (Brink *et al.* 1990). Thus there does not appear to be a host-symbiont specific O-antigen structure.

Changes in the Structure of Rhizobium *LPSs as a Function of Symbiotic Infection*

Using monoclonal antibodies to *R. leguminosarum* bv. viciae bacteroids, Brewin and coworkers have shown that there are changes in LPS epitopes which occur during differentiation from bacteria to bacteroids (Brewin *et al.* 1986; Kannenberg & Brewin, 1989; Wood *et al.* 1989; Sindhu *et al.* 1990; Brewin *et al.* 1990b). Some of these changes can be produced outside the nodule by growing the bacteria at low pH, low O_2 tension, or by using succinate rather than glucose as a carbon source (Kannenberg & Brewin, 1989; Sindhu *et al.* 1990). These epitope changes occur only in the complete LPS molecule which contains all three LPS structural regions; *i.e.* lipid A, core and O-antigen polysaccharide. Recently mutants have been developed which are unable to undergo these epitope changes. Some of these mutants are defective in that the resulting nodules are ineffective (Brewin *et al.* 1990a). Similar results have been reported by Noel and coworkers for *R. leguminosarum* bv. phaseoli mutants (Tao & Noel, 1990). In this case, monoclonal antibodies to bacteria were used and it was found that two LPS epitopes disappear during nodulation. Again, these epitope changes are observed only in the complete from of the LPS. Some of these epitope changes also occur when the bacteria are grown at low pH (Tao & Noel, 1990). Preliminary results indicate that the glycosyl compositions of LPSs from pH 7 and pH 5 grown bacteria show that there is a change in methylated glycosyl residues, 2,3,4-tri-O-methylfucose disappears and 2,3-di-O-methylfucose appears (Bhat and Carlson, unpublished). These results are being investigated further. Noel and coworkers have also observed that one of these epitope changes is dependent on the presence a gene(s) in the *nod* region of the symbiotic plasmid and is controlled by bean root exudate (Tao & Noel, 1990). Thus it appears that subtle changes in the complete LPS molecule may be essential for differentiation of a bacterium to a nitrogen-fixing bacteroid. Since these changes occur in the complete form of the LPS, it may be presumed that the changes are dependent on the presence of the O-antigen polysaccharide portion of the LPS.

Conclusions

It can be concluded from the reports summarized in this chapter that *Rhizobium* LPSs play an important role(s) in establishing a nitrogen-fixing symbiosis with the host legume. In the case of determinate nodule forming symbioses, this role involves the development of the infection thread; while for indeterminate nodule forming symbioses, the LPS is important in the release of bacteria into the root cortical cells. The exact function(s) of the the LPS in symbiotic infection is not known. One could argue that mutants which have LPSs that lack their O-antigenic polysaccharide are grossly affected in their outer membrane and become more susceptible to the host defense response, and/or stimulate the host defense reponse. However, examination of some of these mutants has shown that there are no obvious changes in the outer membrane protein content, release of outer membrane proteins (Noel *et al.* 1986), or in EPS synthesis and structure. Thus, there do not appear to be gross alterations in the outer membranes of these LPS mutants. The fact that the LPS epitope changes found during symbiosis occur only in the form of the LPS that contains the O-antigenic polysaccharide (a) confirms the importance of the presence of this polysaccharide and (b) suggests that subtle structural alterations occur and are required for symbiosis. In spite of this apparent subtle structural requirement, the overall structure of the O-antigen polysaccharide does not seem to be important since one can replace a bv. trifolii with a bv. phaseoli O-antigen polysaccharide. This may suggest that important LPS structural features involved in symbiosis depend on the site of attachment of the O-antigen to the core region and not on the actual structure of the O-antigen polysaccharide. It is also possible that the complete LPS has a different fatty acylation pattern than the incomplete LPS and it is this fatty acylation pattern which provides the correct LPS structural features required for symbiosis. Further work is in progress in our laboratory to determine the structures of the LPS epitopes that change during, and appear to be required for, symbiosis.

Acknowledgement

Some of the work described in this chapter was supported in part by Public Health Service grant RO1-GM-39583-02 and by the DOE/USDA/NSF Plant Science Center program (funded by grant DE-FG09-87ER13810 from the Department of Energy).

References

Battisti, L., Lee, C.C. & Leigh, J.A. (1990) *MPI* **Interlaken Symposium,** 117-P164 Abs.. (Abstract)

Bhat, U.R. & Carlson, R.W. (1990) *International Endotoxin Society* **Abstract,** I-P-29. (Abstract)

Bhat, U.R., Mayer, H., Yokota, A., Hollingsworth, R.I. & Carlson, R.W. (1991a) *J.Bacteriol.* (In Press)

Bhat, U.R., Bhagyalakshmi, S.K. & Carlson, R.W. (1991b) *Carbohydr.Res.* (In Press)

Bhat, U.R., Carlson, R.W., Busch, M. & Mayer, H. (1991c) *Int.J.Syst.Bacteriol.* (In Press)

Borthakur, D., Barber, C.E., Lamb, J.W., Daniels, M.J., Downie, J.A. & Johnston, A.W.B. (1986) *Mol.Gen.Genet.* **203**, 320-323.

Brewin, N.J., Wood, E.A., Larkins, A.P., Galfre, G. & Butcher, G.W. (1986) *J.Gen.Microbiol.* **132**, 1959-1968.

Brewin, N.J., Rae, A.L., Perotto, S., et al (1990a) In: *Nitrogen fixation: Achievements and objectives. Proceedings of the 8th International Congress on Nitrogen Fixation, Knoxville, Tennessee, U.S.A., May 20-26 1990*, Gresshoff, P.M., Roth, L.E., Stacey, G. & Newton, W.E. (eds.), Chapman and Hall, New York and London, pp 227-234.

Brewin, N.J., Kannenberg, E.L., Perotto, S. & Wood, E.A. (1990b) *MPI* **Interlaken Symposium**, 23-L44 Abs.. (Abstract)

Brink, B.A., Miller, J., Carlson, R.W. & Noel, K.D. (1990) *J.Bacteriol.* **172**, 548-555.

Caetano-Anolles, G.C. & Favelukes, G. (1986) *Appl.Environ.Microbiol.* **52**, 377-382.

Carlson, R.W., Sanders, R.E., Napoli, C. & Albersheim, P. (1978) *Plant Physiol.* **62**, 912-917.

Carlson, R.W. (1982) In: *Nitrogen Fixation, Vol. 2*, Rhizobium, Broughton, W.J. (ed.), Clarendon Press, Oxford, pp 199-234.

Carlson, R.W. (1984) *J.Bacteriol.* **158**, 1012-1017.

Carlson, R.W. & Lee, R.P. (1983) *PlantPhysiol.* **71**, 223-228.

Carlson, R.W., Garcia, F., Noel, K.D. & Hollingsworth, R.I. (1990) *Carbohydr.Res.* **195**, 101-110.

Carlson, R.W. & Yadav, M. (1985) *Appl.Environ.Microbiol.* **50**, 1219-1224.

Carlson, R.W., Shatters, R., Duh, J., Turnbull, E., Hanley, B., Rolfe, B.G. & Djordjevic, M.A. (1987) *Plant Physiol.* **84**, 421-427.

Carlson, R.W., Hollingsworth, R.L. & Dazzo, F.B. (1988) *Carbohydr.Res.* **176**, 127-135.

Carlson, R.W., Shatters, R., Duh, J.L., Turnbull, E., Hanley, B., Rolfe, B.G. & Djordjevic, M.A. (1987a) *PlantPhysiol.* **84**, 421-427.

Carlson, R.W., Kalembasa, S., Turowski, D., Pachori, P. & Noel, K.D. (1987b) *J.Bacteriol.* **169**, 4923-4928.

Carrion, M., Bhat, U.R., Reuhs, B. & Carlson, R.W. (1990) *J.Bacteriol.* **172**, 1725-1731.

Chen, H., Batley, M., Redmond, J. & Rolfe, B.G. (1985) *J.Plant Physiol.* **120**, 331.

Diebold, R. & Noel, K.D. (1989) *J.Bacteriol.* **171**, 4821-4830.

Djordjevic, S.P., Chen, H., Batley, M., Redmond, J.W. & Rolfe, B.G. (1987) *J.Bacteriol.* **169**, 53-60.

Dreyfus, B., Garcia, J.L. & Gillis, M. (1988) *Int.J.Syst.Bacteriol.* **38**, 89-98.

Gamo, T., Oyaizu, H. & Mori, K. (1990) In: *Nitrogen fixation: Achievements and objectives. Proceedings of the 8th International Congress on Nitrogen Fixation, Knoxville, Tennessee, U.S.A., May 20-26, 1990*, Gresshoff, P.M., Roth, L.E., Stacey, G. & Newton, W.E. (eds.), Chapman and Hall, New York and London, pp 827-827.

Gray, J.X. & Rolfe, B.G. (1990) *Mol.Microbiol.* **4**, 1425-1431.

Hitchcock, P.J. & Brown, T.M. (1983) *J.Bacteriol.* **154**, 269-277.

Hollingsworth, R.I., Carlson, R.W., Garcia, F. & Gage, D.A. (1989a) *J.Biol.Chem.* **264**, 9294-9299.

Hollingsworth, R.I. & Carlson, R.W. (1989b) *J.Biol.Chem.* **264**, 9300-9303.

Hollingsworth, R.I., Carlson, R.W., Garcia, F. & Gage, D.A. (1989c) *J.Biol.Chem.* **264**, 9294-9299.

Hollingsworth, R.I. & Lill-Elghanian, D.A. (1989d) *J.Biol.Chem.* **264**, 14039-14042.

Hotter, G.S. & Scott, D.B. (1991) *J.Bacteriol.* **173**, 851-859.

Kannenberg, E.L. & Brewin, N.J. (1989) *J.Bacteriol.* **171**, 4543-4548.

Kneen, B.E., LaRue, T.A., Hirsch, A.M., Smith, C.A. & Weeden, N.F. (1990) *Plant Physiol.* **94**, 899-905.

Krauss, J.H., Weckesser, J. & Mayer, H. (1988) *Int.J.Syst.Bacteriol.* **38**, 157-163.

Kreutzer, d.L., Bulle, C.S. & Robertson, D.C. (1979) *Infect.Immun.* **23**, 811-819.

Lagares, A. & Favelukes, G. (1988) In: *Nitrogen Fixation: Hundred Years After*, Bothe, De Bruijn & Newton (eds.), Gustav Fischer, Stuttgart and New York, pp 476.

Law, I.J., Yamamoto, Y., Mort, A.J. & Bauer, W.D. (1982) *Planta* **154**, 100-109.

Marx, A., Ionescu, J. & Pop, A. (1983) *Zentralbl.Bakteriol.Parasitenkd.Infektionskr.Hyg.Abt.J.* 758-763.

Noel, K.D., VandenBosch, K.A. & Kulpaca, B. (1986) *J.Bacteriol.* **168**, 1392-1401.

Noel, K.D. (1991) In: *Molecular signals in plant-microbe communication*, Verma, D.P. (ed.), CRC Press, Boca Raton, pp in press.

Peterson, A.A. & McGroarty, E.J. (1985) *J.Bacteriol.* **162**, 738-745.

Priefer, U.B. (1989) *J.Bacteriol.* **171**, 6161-6168.

Putnoky, P., Grosskopf, E., Ha, D.T.C., Kiss, G.B. & Kondorosi, A. (1988) *J.Cell Biol.* **106**, 597-607.

Putnoky, P., Petrovics, G., Kereszt, A., Grosskopf, E., Ha, D.T.C., Bánfalvi, Z. & Kondorosi, A. (1990) *J.Bacteriol.* **172**, 5450-5458.

Puvanesarajah, V., Schell, F.M., Gerhold, D. & Stacey, G. (1987) *J.Bacteriol.* **169**, 137-141.

Ryan, J.M. & Conrad, H.E. (1974) *Arch.Biochem.Biophys.* **162**, 530-535.

Sindhu, S.S., Brewin, N.J. & Kannenberg, E.L. (1990) *J.Bacteriol.* **172**, 1804-1813.

Stacey, G., So, J.-S., Roth, L.E., Bhagya Lakshmi, S.K. & Carlson, R.W. (1991) *MPMI* (In Press)

Tao, H. & Noel, D. (1990) *Nitrogen Fixation Congress* **Knoxville, TN**, (Abstract)

Urbanik-Sypniewska, T., Seydel, U., Greck, M., Weckesser, J. & Mayer, H. (1989) *Arch.Microbiol.* **152**, 527-532.

Westphal, O. & Jann, K. (1965) *Meth.Carbohydr.Chem.* **5**, 83-91.

Whittenburry, R. & McLee, G.A. (1967) *Arch.Microbiol.* **59**, 324-334.

Williams, M.N.V., Hollingsworth, R.I., Klein, S. & Signer, E.R. (1990a) *J.Bacteriol.* **172**, 2622-2632.

Williams, M.N.V., Hollingsworth, R.I., Brzoska, P.M. & Signer, E.R. (1990b) *J.Bacteriol.* **172**, 6596-6598.

Wood, E.A., Butcher, G.W., Brewin, N.J. & Kannenberg, E.L. (1989) *J.Bacteriol.* **171**, 4549-4555.

Zevenhuizen, L.P.T.M., Posthumus, M.A. & Scholten-Koerselman, H.J. (1980) *Arch.Microbiol.* **125**, 1-8.

Molecular signaling in the *Bradyrhizobium japonicum*-soybean symbiosis

Gary Stacey

Center for Legume Research, Department of Microbiology and Graduate Program of Ecology, The University of Tennessee, Knoxville, TN 37996-0845, USA

Introduction

Close associations (e.g., pathogenic or mutualistic) between bacteria and eukaryotes require close coordination and integration of a variety of cellular processes. Recent evidence indicates that reciprocal exchange of diffusible signal molecules represents a kind of "communication" that is important for the establishment of these associations. The definition of these communication networks can lead to strategies to enhance or terminate the interaction. Perhaps the best studied example of such communication is the symbiotic interaction between legumes and rhizobia. *Rhizobium, Bradyrhizobium*, and *Azorhizobium* species are Gram⁻ bacteria that fix nitrogen in a symbiotic association with leguminous plants. These symbioses are valuable models for the study of bacterial-eukaryotic interactions. Recent research has identified signal molecules that are exchanged between the plant and bacteria that serve to induce transcription of genes essential for establishment of the symbiosis. One example is flavonoids exuded by the plant and recognized by the bacteria; these flavonoids are the inducers of the *nod* genes that are essential for infection (nodulation) of the plant (reviewed in Long, 1989). It has now been shown that these *nod* genes encode products that are involved in the synthesis of additional signal molecules that are recognized by the plant. These bacterial products lead to root hair curling and plant cortical cell division, two of the initial, essential steps in the infection process (Kijne, 1991; Roth & Stacey, 1991).

nod gene regulation

Figure 1 displays the current genetic map of *Bradyrhizobium japonicum*, symbiont of soybean, highlighting symbiotically important genes. The *fix* and *nif* genes are involved either directly in the nitrogen fixation process or have been found to be important for the maintenance of the symbiotic state. The *nod* (and *nol*) genes encode products that are essential for the establishment of the symbiosis. The *nod* genes can be placed into two general groups; the "common" *nod* genes, e.g., *nodDABCIJ*, and the host-specific *nod* genes, e.g., *nodZ* and *nolA* in *B. japonicum*.

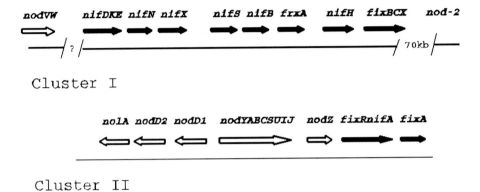

Figure 1. Current genetic map of Bradyrhizobium japonicum USDA110 highlighting symbiotic genes. Arrows indicate direction of transcription. open arrows: nodulation genes; closed arrows: nitrogen fixation genes.

The former group contains genes that are found in all rhizobia and appear to be similar both in sequence and function. The latter genes are involved in determining a specific host range for each rhizobia. For example, the *nodQPH* genes are involved in specifying that *R. meliloti* can nodulate alfalfa (Faucher et al, 1988,1989a, 1989b; Lerouge et al,1990b; Schwedock & Long, 1990). In the case of *B. japonicum, nodZ* is essential for nodulation of siratro (Nieuwkoop et al,1987; Stacey et al,1989), an alternative host, while *nolA* determines the ability to nodulate a certain subset of soybean cultivars (Sadowsky et al,1991). The *nodD* gene is a common *nod* gene, but also has host-specific character. The *nodD* protein product is a positive activator required for transcription of most of the other *nod* genes (reviewed in Long, 1989). The current model stipulates that NodD recognizes host-produced flavonoids and stimulates the transcription of the *nod* genes required for infection. This is true of all rhizobia that have been examined. NodD also acts in a host-specific fashion in that each rhizobia recognizes specifically the flavonoids produced by its host species (Long, 1989). For example, *B. japonicum* NodD recognizes isoflavones, e.g., genistein and daidzein, exuded by soybean, but does not activate *nod* gene expression if the flavone luteolin is added (Kosslak et al,1987; Banfalvi et al,1988; Gottfert et al,1988). Luteolin is a good inducer of *nod* gene expression in *R. meliloti* (Mulligan & Long, 1985). NodD activates transcription by binding to a conserved DNA sequence, the *nod* box, found 5' of each *nod* operon (Hong et al,1987; Fisher & Long, 1989; Kondorosi et al,1989).

A. Practical application of information concerning chemo-signaling pathways

As indicated above, one purpose of elucidating the chemo-signaling pathway between host and symbiont is to allow manipulation to enhance the symbiosis. For example, a significant problem for the introduction of superior *B. japonicum* strains for soybean production has been competition from inferior, indigenous strains in the soil (e.g., Caldwell & Vest, 1970). When new inoculant strains are tested, they are usually found to occupy only a few percent of the nodules formed on soybean plants grown under production conditions. In order to overcome this problem of competition by soil rhizobia, we sought to exploit the chemical signaling network between plant and symbiont to enhance the competitiveness of inoculant strains (Cunningham et al,1991). The specific goal was to develop a combination of a highly efficient *B. japonicum* strain which, when applied in the presence of a particular chemical, would out-compete the indigenous population in the formation of nodules on soybean. Approximately 1000 structural and functional analogs of known inducers or inhibitors of *B. japonicum nod* gene expression were tested. These compounds were screened using *B. japonicum* strains containing an inducible *nodY-lacZ* fusion as an assay system (Banfalvi et al,1988). Figure 2 shows the results with 7-hydroxy-5-methylflavone, the best inhibitor found (Cunningham et al,1991). This compound was a strong inhibitor of *nod* gene expression in strain USDA110 and mildly inhibitory to strain USDA31. The inclusion of this inhibitor in a competition study between these two strains significantly reduced nodulation by USDA110 to the point that virtually all the nodules formed (96%) contained USDA31. Although this experiment definitely indicated that chemical treatments based on *nod* gene induction could be used to modulate interstrain competition, the conclusion of our study was that this method would be extremely difficult to adapt for practical use. The major problem was that, although 7-hydroxy-5-methylflavone was a good inhibitor of 5 strains of *B.japonicum* that were tested, it was a strong inducer of strain USDA3. Therefore, it appeared unlikely that one compound could be found that would act to inhibit all of the indigenous strains that would be encountered in a field situation. Therefore, given this unsuspected variability, the cost of registration of an agronomic chemical, as well as the potential for developing resistant field populations, it was judged that this approach could not be developed for practical use. However, for the purposes of this article, the example of 7-hydroxy-5-methylflavone adequately demonstrates that knowledge of the chemo-signaling pathways used between bacteria and their partners can be used to modify these associations.

B. *nodD* regulation in *B. japonicum*

As indicated above, NodD activates transcription of the other *nod* genes in response to the presence of plant-produced flavonoids. In this case, the same flavonoid inducer activates all of the *nod* genes in any given *Rhizobium* species (reviewed in Long, 1989; Roth & Stacey, 1991). In the majority of the systems studied, *nodD* is transcribed constitutively (Long, 1989; Roth & Stacey, 1991). Therefore, it was

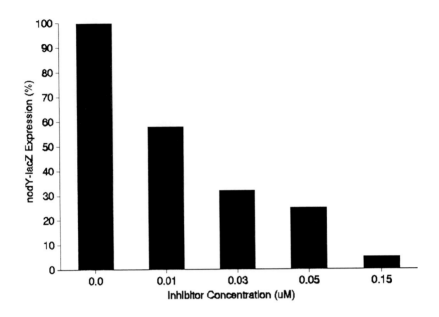

Figure 2. Effect of 7-hydroxy-5-methylflavone on nodY-lacZ *expression in* B. japonicum *in the presence of 3μM genestein. Modified from Cunningham et al. (1991)*

surprising to find that the *nodD* of B. *japonicum* was inducible by the same flavonoids that induce the other *nod* genes (Banfalvi et al,1988). Indeed, NodD is autoregulatory since it is required for *nodD* transcription. Another interesting observation concerning *nod* gene expression in B. *japonicum* is that the natural, plant-derived sources of flavonoids (e.g., soybean seed and root extracts) always gave significantly higher induction than any of the purified inducers (e.g., genistein or daidzein). This suggested that the natural sources may possess special properties. We examined this possibility by analyzing soybean seed and root extracts by HPLC and identified several flavonoid compounds (Smit et al,1990). Only two compounds, genistein and daidzein, found in these extracts were strong inducers of *nodD* and other *nod* genes. However, several purified compounds could specifically induce *nodD* expression. These compounds were identified as glycosylated derivatives of genistein or daidzein (e.g., 6"-malonylgenistin). These results offer an explanation as to why the natural sources of flavonoids gave enhanced induction ability. That is, these sources contain compounds that specifically induce *nodD* expression; the resulting increased level of NodD in the cell leads to enhanced expression of the other *nod* gene operons.

NodD activation of *nod* gene transcription requires binding to the *nod* box promoter sequence (Hong et al,1987; Fisher & Long, 1989; Kondorosi et al,1989). A puzzling aspect of *nodD* induction in *B. japonicum* was the apparent absence of a sequence showing significant homology to the *nod* box. However, a divergent, *nod*-box like sequence was identified 5' of the *nodD* gene and it was proposed that this sequence was a functional *nod* box (Nieuwkoop et al,1987). Recently, Wang and Stacey (1991) have shown that deletion of this sequence eliminates *nodD* induction; therefore, it is likely that this sequence is a functional *nod* box. However, these results raised questions about the sequence conservation exhibited by *nod* boxes in various *Rhizobium*, *Azorhizobium*, and *Bradyrhizobium* species. Therefore, an examination of these sequences was undertaken to look for a structure that would explain the *B. japonicum nod* box, as well as functionally related sequences from other rhizobia. This investigation led to the proposal that the *nod* box is composed of a repeated 9 bp sequence (Figure 3). Most *nod* boxes that have been sequenced possess 4 such repeats separated by 4 bp. Insertion mutations within this 4 bp gap show that the orientation of the two 9 bp dimers on opposite sides of the DNA helix is essential for optimal *nod* gene expression (Wang & Stacey, 1991). Therefore, these dimers do appear to be functional units. In the case of the *nodD nod* box in *B. japonicum*, only two 9 bp repeats are present, but this appears sufficient to allow NodD-dependent induction. In addition, it is quite possible that the differences between the *nodD nod* box and those found 5' of the other *nod* gene operons of *B. japonicum* may play a role in the specific induction of *nodD* by glycosylated genistein and daidzein analogs. We are currently investigating this latter possibility.

nod gene function

Evidence is now accumulating to suggest that many of the *nod* genes, if not all, are involved in the production of a "phytohormone" substance which induces root hair curling and cortical cell division. Although numerous earlier reports had reported the isolation of material from rhizobial cultures that could affect leguminous roots, the first indication of the involvement of the *nod* genes in this process came from the work of van Brussel and colleagues (van Brussel et al,1986; Zaat et al,1987; van Brussel et al,1990). These authors reported that sterile culture supernatants from *R. leguminosarum* bv. *viciae* cultures would elicit a thick and short root (TSR) phenotype on seedlings of common vetch (*Vicia sativa* subsp. *nigra*). This work also showed that the *nodD* and *nodABC* genes were essential for this effect and that inducers of the *nod* genes must be present. Subsequently, Schmidt et al. (1988) reported that culture supernatants from luteolin-induced *R. meliloti* cells would induce cell division in soybean and alfalfa protoplasts. Banfalvi and Kondorosi (1989) reported that expression of the *nodABC* genes of *R. meliloti* in *E. coli* resulted in the release of substances into the culture supernatant which could induce root hair curling on alfalfa and a few other plants. The compounds released by *Rhizobium* appeared to be specific for their host plant. Faucher et al. (1988,1989b) reported that, although culture supernatants from *R. meliloti* cells would curl root hairs on a wide variety of plants, the addition of the *nodH* gene rendered the supernatants specific for alfalfa. In addition, mutants in the *nodQ* gene produced compounds which could curl the root hairs of vetch and alfalfa (Banfalvi & Kondorosi, 1989; Cervantes et al,1989). At least three bioactive compounds were

isolated by HPLC from culture supernatants of *R. meliloti* and when added
separately could induce root hair curling or when added together could induce
significant root cortical cell division to yield an empty, nodule-like structure
(Faucher et al,1989b). The structure of two of the compounds produced by *R. meliloti*
has been determined. The first of these compounds, NodRm-1 is a N-acyl tri N-
acetyl-1,4-D-glucosamine tetrasaccharide, bearing a sulfate group on C-6 of the
reducing sugar (Lerouge et al,1990b, Figure 4). The second compound isolated,
NodRm-2, is identical to NodRm-1, but lacks the sulfate substituent. It now seems
likely that the NodH, NodP, and NodQ proteins are involved in the conversion of
NodRm-1 to NodRm-2 via a sulfation reaction (Lerouge et al,1990b; Schwedock &
Long, 1990). This modification is apparently important to the specificity of this
compound for *Medicago*. A similar compound has very recently been identified
from culture supernatants of *R. leguminosarum* bv. *viciae*; however, it is a
pentasaccharide (Spaink et al,1991).

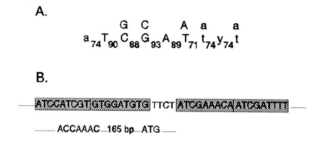

Figure 3. Repetitive model for the nod *box. A: the 9 bp repeat found in the* nod *box.*
Subscripts give the percent occurrence within 72 repeats examined from various nod *boxes.*
Capital letters identify the most conserved bases. B: the nod *box found 5' of the* nodY *in B.*
japonicum. *The 9 bp repeats are shaded. Modified from Wang and Stacey (1991).*

These compounds either in the crude or purified state have been shown to have a
variety of biological activities: 1. They can inhibit root growth (the TSR phenotype)
(van Brussel et al,1986; Zaat et al,1987). 2. They can induce plant root hair
deformation (Had[+] phenotype, Lerouge et al,1990a,b; Banfalvi & Kondorosi, 1989). 3.
They can induce cortical cell division (Coi[+] phenotype, Lerouge et al,1990a,b). 4.
They can induce nodule formation (Nod[+] phenotype, Lerouge et al,1990a; Spaink et
al,1991). Of course, these nodule structures lack bacteria. 5. They can induce
increased exudation of flavonoids, the inducers of *nod* gene expression (the INI,
increased *nod* gene induction, phenotype, van Brussel et al,1990). These biological
effects of the isolated *nod* factor mirror various aspects of the bacterial infection
process. Therefore, it now seems likely that these plant responses to the bacteria are
due to interaction with the *nod* factor(s).

NodRm-1

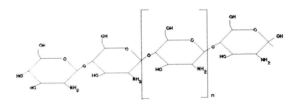

Oligochitosan

Figure 4. Two examples of oligosaccharins. NodRm-1 is the nod factor produced by R. meliloti (LeRouge et al. 1990b). Oligochitosan is a known elicitor of lignification (Barber et al. 1989).

We have recently used the method of Spaink et al. (1991) to analyze *B. japonicum* cultures for the production of *nod* factors such as those described above from *R. meliloti*. This method involves labeling flavonoid-induced cultures with ^{14}C-acetate and subsequent analysis of butanol-extracted material by thin-layer-chromatography. The various spots corresponding to the labeled material can then be visualized by autoradiography. Using this method, we have detected labeled compounds that are specific to genistein or seed extract-induced cultures. These compounds are candidates for the *nod* factors of *B. japonicum*. Furthermore, induced cultures from NodD⁻, NodC⁻, or NodA⁻ *B. japonicum* strains do not produce these compounds; therefore, suggesting that these genes are essential for their production. Further analysis will be required to confirm that the compounds tested are functionally or structurally related to the *nod* metabolites isolated from *R. meliloti* or *R. leguminosarum* bv. *viciae*. However, the crude extracts containing the ^{14}C-labeled compounds do induce a TSR response on soybean and will also induce the curling of root hair cells (Sanjuan and Stacey, unpubl.). Therefore, it is likely that *B. japonicum* will be found to produce lipo-oligosaccharide molecules essential for nodulation.

The fact that an oligosaccharide is involved in the induction of nodule formation has precedence in the literature. In fact, a class of molecules called oligosaccharins (oligosaccharides with growth regulatory properties) are known plant signal molecules (Albersheim et al,1983; Albersheim & Darvill, 1985; Hahn & Cheong, 1991). These oligosaccharins have been characterized from plant, fungal, and bacterial sources (Hahn et al,1981; Nothnagel et al,1983; York et al,1984; McDougall & Fry, 1988; Barber et al,1989; Lerouge et al,1990b). Various oligosaccharins have been shown to activate the plant defense response during infection (Darvill & Albersheim, 1984; Ryan, 1987; Hahn et al,1989), to regulate plant hormone responses (York et al,1984; McDougall & Fry, 1988, 1989a, 1989b), and affect plant development (Eberhard et al,1989). Therefore, it is perhaps not surprising that a rhizobial oligosaccharin, the *nod* factor, is involved in determining symbiosis since this process requires cell-cell recognition and regulation of a host cellular response. Comparison of the *nod* factors with known oligosaccharins (e.g., Figure 4) shows they are most closely related to the biologically active chitin oligomers. In fact, removal of the sulfate and fatty acyl group from the *R. meliloti* NodRm-1 compound would be expected to produce a chitin tetrasaccharide which is known to be very active in eliciting a defense response (e.g. lignification, Figure 4, Barber et al,1989). If one views a *Rhizobium*-legume symbiosis as a modified pathogen-plant interaction, the host-specific *nod* genes, and possibly the common *nod* genes, can be viewed as analogous to bacterial avirulence genes which modify a chitin-like molecule that would otherwise be recognized by the plant resistance gene product. The fact that the *nod* genes are induced by molecules, flavonoids, from the host which are intermediates in the plant defense pathway leading to phytoalexin production make this comparison to plant pathogens even more interesting.

Summary

Although our current knowledge concerning the signaling between rhizobia and their leguminous plant hosts is incomplete, a clear picture is emerging of the reciprocal exchange of signal molecules. The plant exudes specific flavonoids that serve to induce *nod* gene expression in the bacteria. These *nod* genes encode proteins that are involved in the production of the complementary signal molecule that elicits the necessary host response for nodulation. One example is given above as to how such knowledge about signaling between the two symbiotic partners can be used to manipulate this interaction for practical benefit. However, perhaps more exciting, is the basic information that these studies provide concerning cellular signal transduction mechanisms in bacteria and plants. The discovery of novel, plant growth regulatory molecules, *nod* factors, produced by rhizobia opens up new opportunities to study plant recognition and cellular signaling mechanisms. Therefore, further investigation of rhizobia-legume interactions may well lead to discoveries with significance far beyond symbiotic nitrogen fixation.

Acknowledgement

Work from the author's laboratory was supported by grants from the National Institutes of Health and from the Competitive Grants Program of the U.S. Department of Agriculture. Special thanks to Russell Carlson, Scott Cunningham, Herman Spaink, Gerrit Smit, and Shui-Ping Wang for stimulating discussions.

References

Albersheim, P., Darvill, A.G., McNeil, M., Valent, B.S., Sharp, J.K., Nothnagel, E.A., Davis, K.R., Yamazaki, N., Gollin, D.J., York, W.S., Dudman, W.F., Darvill, J.E. & Dell, A. (1983) in Structure and function of plant genomes, eds. Ciferri, O. & Dure, L.III (Plenum Publ. Corp., New York, N.Y.), pp. 293-312.

Albersheim, P. & Darvill, A.G. (1985) Sci. Am. 253, 58-64.

Banfalvi, Z., Nieuwkoop, A.J., Schell, M.G., Besl, L. & Stacey, G. (1988) Mol. Gen. Genet. 214, 420-424.

Banfalvi, Z. & Kondorosi, A. (1989) Plant Mol. Biol. 13, 1-12.

Barber, M.S., Bertram, R.E. & Ride, J.P. (1989) Physiol. Mol. Plant Pathol. 34, 3-12.

Caldwell, B.E. & Vest, G. (1970) Crop Sci. 10, 19-21.

Cervantes, E., Sharma, S.B., Maillet, F., Vasse, J., Truchet, G. & Rosenberg, C. (1989) Mol. Microbiol. 3, 745-755.

Cunningham, S., Kollmeyer, W.D. & Stacey, G. (1991) Appl. Environ. Microbiol. (in press)

Darvill, A.G. & Albersheim, P. (1984) Ann. Rev. Plant Physiol. 35, 243-275.

Eberhard, S., Doubrava, N., Marfa, V., Mohnen, D., Southwick, A., Darvill, A. & Albersheim, P. (1989) Plant Cell 1, 747-755.

Faucher, C., Maillet, F., Vasse, J., Rosenberg, C., van Brussel, A.A.N., Truchet, G. & Denarie, J. (1988) J. Bacteriol. 170, 5489-5499.

Faucher, C., Lerouge, P., Roche, P., Rosenberg, C., Debelle, F., Vasse, J., Cervantes, E., Sharma, S.B., Truchet, G., Prome, J.-C. & Denarie, J. (1989a) in Signal Molecules in Plants and Plant-Microbe Interactions, ed. Lugtenberg, B.J.J. (Springer-Verlag, Berlin), pp. 379-386.

Faucher, C., Camut, S., Denarie, J. & Truchet, G. (1989b) Mol. Plant-Microbe Int. 2, 291-300.

Fisher, R.F. & Long, S.R. (1989) J. Bacteriol. 171, 5492-5502.

Gottfert, M., Weber, J. & Hennecke, H. (1988) J. Plant Physiol. 132, 394-397.

Hahn, M.G., Darvill, A.G. & Albersheim, P. (1981) Plant Physiol. 68, 1161-1169.

Hahn, M.G., Bucheli, P., Cervone, F., Doares, S.H., O'Neill, R.A., Darvill, A. & Albersheim, P. (1989) in Plant-microbe interactions. Molecular and genetic perspectives, Vol. 3, eds. Kosuge, T. & Nester, E.W. (McGraw Hill Publ. Co., New York, N.Y.), pp. 131-181.

Hahn, M.G. & Cheong, J.-J. (1991) in Advances in molecular genetics of plant-microbe interactions, eds. Hennecke, H. & Verma, D.P.S. (Kluwer Acad. Publ., Dordrecht), pp. 403-420.

Hong, G.-F., Burn, J.E. & Johnston, A.W.B. (1987) Nucl. Acids Res. 15, 9677-9689.

Kijne, J. (1991) in Biological Nitrogen Fixation, eds. Stacey, G., Burris, R.H. & Evans, H.J. (Chapman and Hall, New York), (in press).

Kondorosi, E., Gyuris, J., Schmidt, J., John, M., Duda, E., Hoffman, B., Schell, J. & Kondorosi, A. (1989) EMBO J. 8, 1331-1340.

Kosslak, R.M., Bookland, R., Barkei, J., Paaren, H.E. & Appelbaum, E.R. (1987) Proc. Natl. Acad. Sci. (USA) 84, 7428-7432.

Lerouge, P., Roche, P., Prome, J.-C., Faucher, C., Vasse, J., Maillet, F., Camut, S., de Billy, F., Barker, D.G., Denarie, J. & Truchet, G. (1990a) in Nitrogen Fixation: Achievements and Objectives, eds. Gresshoff, P.M., Roth, L.E., Stacey, G. & Newton, W.E. (Chapman and Hall, New York), pp. 177-186.

Lerouge, P., Roche, P., Faucher, C., Maillet, F., Truchet, G., Prome, J.C. & Denarie, J. (1990b) Nature 344, 781-784.

Long, S.R. (1989) Cell 56, 203-214.

McDougall, G.J. & Fry, S.C. (1988) Planta 175, 412-416.

McDougall, G.J. & Fry, S.C. (1989a) Plant Physiol. 89, 883-887.

McDougall, G.J. & Fry, S.C. (1989b) J. Exp. Bot. 40, 233-238.

Mulligan, J.T. & Long, S.R. (1985) Proc. Natl. Acad. Sci. (USA) 82, 6609-6613.

Nieuwkoop, A.J., Banfalvi, Z., Deshmane, N., Gerhold, D., Schell, M.G., Sirotkin, K.M. & Stacey, G. (1987) J. Bacteriol. 169, 2631-2638.

Nothnagel, E.A., McNeil, M., Albersheim, P. & Dell, A. (1983) Plant Physiol. 71, 916-926.

Roth, L.E. & Stacey, G. (1991) in Microbial Cell-Cell Interactions, ed. Dworkin, M. (ASM Press, Washington, D.C.), pp. in press.

Ryan, C.A. (1987) Ann. Rev. Cell Biol. 3, 295-317.

Sadowsky, M.J., Cregan, P.B., Gottfert, M., Sharma, A., Gerhold, D., Rodriguez-Quinones, F., Keyser, H.H., Hennecke, H. & Stacey, G. (1991) Proc. Natl. Acad. Sci. (USA) 88, 637-641.

Schmidt, J., Wingender, R., John, M., Wieneke, U. & Schell, J. (1988) Proc. Natl. Acad. Sci. (USA) 85, 8578-8582.

Schwedock, J. & Long, S.R. (1990) Nature 348, 644-646.

Smit, G., Puvanesarajah, V., Carlson, R.W. & Stacey, G. (1990) in Nitrogen Fixation: Achievements and Objectives, eds. Gresshoff, P., Roth, E., Stacey, G. & Newton, W.E. (Chapman and Hall, New York), pp. 274.

Spaink, H.P., Geiger, O., Sheeley, D.M., van Brussel, A.A.N., York, W.S., Reinhold, V.N., Lugtenberg, B.J.J. & Kennedy, E.P. (1991) in Advances in Molecular Genetics of Plant-Microbe Interactions, eds. Hennecke, H. & Verma, D.P.S. (Kluwer Acad. Publ., Dordrecht), 142-149.

Stacey, G., Schell, M.G. & Deshmane, N. (1989) in Signal Molecules in Plants and Plant-Microbe Interactions, ed. Lugtenberg, B.J.J. (Springer-Verlag, Berlin), pp. 394-399.

van Brussel, A.A.N., Zaat, S.A.J., Canter-Cremers, H.C.J., Wijffelman, C.A., Pees, E., Tak, T. & Lugtenberg, B.J.J. (1986) J. Bacteriol. 165, 517-522.

van Brussel, A.A.N., Recourt, K., Pees, E., Spaink, H.P., Tak, T., Wijffelman, C.A., Kijne, J.W. & Lugtenberg, B.J.J. (1990) J. Bacteriol. 172, 5394-5401.

Wang, S.-P. & Stacey, G. (1991) J. Bacteriol. 173, in press.

York, W.S., Darvill, A.G. & Albersheim, P. (1984) Plant Physiol. 75, 295-297.

Zaat, S.A.J., van Brussel, A.A.N., Tak, T., Pees, E. & Lugtenberg, B.J.J. (1987) J. Bacteriol. 169, 3388-3391.

Cytokinins and legume nodulation

Barbara J. Taller

Department of Biology, Memphis State University, Memphis, TN 38152, USA

Introduction

Many microorganisms involved in symbiotic or parasitic relationships with plants are characterized by the production of plant hormones (Greene, 1980). Among these compounds are the cytokinins, N^6-substituted adenine derivatives which regulate plant growth and development. While cytokinin synthesis is associated with virulence in some plant pathogens, its role in symbiotic relationships has not been established. There is considerable evidence for cytokinin production by rhizobia and for the involvement of cytokinins in legume nodulation. However, neither rhizobial cytokinin synthesis nor its role in the legume-rhizobia symbiosis has been well characterized.

Although cytokinins are involved in essentially all aspects of plant development, they are characterized by their ability to promote cell division in plant tissues. The cytokinin isopentenyladenine (iP) is the archetype molecule to which all other naturally-occurring cytokinins are structurally related. The first step in cytokinin biosynthesis is carried out by isopentenyltransferase, which transfers the isoprenoid chain from dimethyl- allylpyrophosphate to 5'AMP. The nucleotide side chain may be hydroxylated to produce the zeatin-type cytokinins, characteristic of plants and many plant-associated microbes.

More than thirty years ago, Skoog and coworkers suggested that the action of rhizobia in nodule formation likely involved the supply or activation of several different growth factors (Arora *et al.*, 1959). Although the involvement of phytohormones in nodulation has been assumed for many years, there is little direct evidence in support of this. Considerable evidence indicates that nodule development involves a perturbation of the root phytohormone balance. Application of cytokinin (Arora *et al.*, 1959; Rodriguez-Barrueco *et al.*, 1973), auxin or auxin-like substances (Allen *et al.*, 1953) can induce pseudonodules on leguminous and non-leguminous plants. Cortical cell division and the nodulin ENOD2 are induced in legume roots in response to the cytokinin benzyladenine (Bauer *et al.*, 1985; Nirunsuksiri and Sengupta-Gopalan, 1988) as well as to auxin transport inhibitors (Hirsch *et al.*, 1989). Cooper and Long (1988) showed that when alfalfa roots were inoculated with *E. coli* containing the *trans*-zeatin secretion (*tzs*) gene from *Agrobacterium*, cortical cell divisions which mimic nodule initiation were induced. More complete development of nodules occurred when nodulation (*nod*) genes were present, suggesting that they might provide a second substance which may act in concert with cytokinins. Similarly, a *Rhizobium* clone containing *tzs* complemented some Nod-minus mutants, leading to nodule-like structures on alfalfa roots (Long and Cooper, 1988).

Cytokinin production by rhizobia

As the literature concerning cytokinin synthesis by rhizobia is fragmentary and contradictory (Morris, 1986; Sturtevant and Taller, 1989), it was of interest to examine cytokinin production. The cytokinin content of culture filtrates was analyzed as described previously (Taller and Sturtevant, 1991). To date, dozens of strains, including the type strains of eight major cross-inoculation groups, have been analyzed (Taller and Sturtevant, 1991). Every strain examined produced at least two cytokinin-active compounds; some as many as four. Most strains secreted zeatin and/or zeatin derivatives. The range of cytokinin activity was equivalent to 0.5 to 4 µg of kinetin per liter of culture, an amount sufficient to promote cell division in tobacco pith cells. A preliminary examination of *Azorhizobium caulinodans* culture filtrate showed it contained over 8 µg kinetin equivalents per liter (Kumar and Taller, unpublished data), perhaps related to its stem nodulation and crack-entry type of infection (de Bruijn, 1989). Despite the occurrence of cytokinins and the relatedness to agrobacteria, isopentenyltransferase activity has not been reported in rhizobia.

There were both quantitative and qualitative differences in cytokinin production between rhizobial strains. None of the *Bradyrhizobium* strains produced iP or its derivatives, while most *Rhizobium* strains did. In contrast, secretion of ribosylzeatin appears characteristic of slow-growing rhizobia, though most fast-growers produced small amounts. The chromatographic properties of a number of these cytokinin-active compounds differed from those cytokinins ascribed to plant-associated microbes. As the diversity of cytokinins known to occur in prokaryotes is considerably less than in plants, the identity of these apparently novel cytokinins is being pursued.

Modification of cytokinin synthesis

The mutual triggering of gene expression seems to be characteristic of many plant-microbe interactions. Plant flavonoids induce bacterial nodulation genes whose products in turn induce plant nodulins. Expression of virulence genes and cytokinin synthesis in *Agrobacterium* is induced by wound-induced compounds such as phenolics and sugars. Experiments were carried out to determine whether rhizobial cytokinin production was influenced by flavonoids (Taller and Sturtevant, 1991). Addition of the appropriate seed extract or flavonoid to the culture medium resulted in a qualitative change in cytokinin synthesis rather than a quantitative one. The increase in one cytokinin was accompanied by a decrease in another compound, the resulting compound always a later-eluting or more lipophilic one. This change was also consistent with a shift to cytokinin species further down the probable biosynthetic pathway, although such pathways have not been clearly established. This primarily qualitative change differs from that in *Agrobacterium* where plant factors cause approximately a ten-fold increase in cytokinin synthesis due to *tzs* expression (Powell *et al.*, 1988). The limited number of *nod* gene inducers which have been examined so far may not optimally induce cytokinin genes in rhizobia.

Adenine, a cytokinin precursor, also affected cytokinin synthesis in defined media. A concentration of 10 mg/L caused a 25% increase in the cytokinin activity of uninduced *R. meliloti* cultures and a two-fold increase in cultures containing 2 μM luteolin. Adenine also stimulated production of an additional cytokinin in the culture medium, which cochromatographed with methylthioisopentenyladenine (Taller and Sturtevant, 1991). Unlike most *Rhizobium* auxotrophs, adenine requiring mutants are symbiotically defective, either Nod- or producing Fix-immature nodules (Pain, 1979 and references therein). This indicates an important role for adenine in nodulation, perhaps as a cytokinin precursor (Pain, 1979).

Genetics of cytokinin production

Determinants of cytokinin synthesis are plasmid borne in *Agrobacterium tumefaciens*, *Pseudomonas savastanoi* and probably *Rhodococcus fascians* (Morris, 1986). Attempts to demonstrate a role for the symbiotic (Sym) plasmid in rhizobial cytokinin synthesis have been unsuccessful. When cytokinin production was compared in pairs of strains, one of which had been cured of the Sym plasmid, no differences were detected. However, this may be caused by gene duplication or by genes remaining silent under noninducing conditions. Nodulation genes do not seem to be required for cytokinin synthesis in the absence of plant flavonoids. Analysis of *nod* gene deletion mutants and transposon mutants showed cytokinin levels similar to the parental strains (Taller and Sturtevant, 1991).

Our limited data concerning the genetics of cytokinin production are consistent with those found for IC3342, an unusual *Rhizobium* which nodulates pigeonpea (*Cajanus cajan* L.) and which overproduces cytokinin (Letham *et al.*, 1990). The high cytokinin levels cause leaf curl syndrome in pigeonpea, characterized by hyponasty, leaf curling, release from apical dominance and reduced internode length. Similar symptoms are seen in tobacco plants transgenic for isopentenyltransferase (*ipt*), which contained 4 times more cytokinin than wild-type plants: reduced height, leaf area, apical dominance and primary root growth, as well as a thicker, shorter root tip region with root hairs closer to the tip (Medford *et al.*, 1989).

Five loci involved in cytokinin synthesis were identified in *Rhizobium* IC3342 (Letham *et al.*, 1990). Two were chromosomal and three were on the Sym plasmid but not at the *nod* loci. Only one of these loci, *lcr* is unique to this strain and thus responsible for the overproduction of cytokinin. This locus did not show sequence homology to the known prokaryotic cytokinin genes *ipt*, *tzs* or *ptz*, but several open reading frames showed homology to *E. coli* two component regulatory genes (Letham *et al.*, 1990). Thus, *lcr* appears to be a unique regulatory locus on the Sym plasmid, not directly responsible for rhizobial cytokinin production.

Rhizobia-legume interactions

As discussed above, application of cytokinin or cytokinin-producing microbes induced cortical cell divisions in legume roots, an early step in nodule development. There is also evidence for the enhancement of nodulation by cytokinin. Growth pouch experiments were carried out in which soybean plants inoculated with *B. japonicum* 61A68 were given a one-time addition of *trans*-zeatin. Plants treated with 0.5, 1 or 5 µg/L zeatin showed twice as many nodules as the control, while 10 µg/L was inhibitory, causing a 50% reduction in nodule number (Taller and Sturtevant, 1991). Enhancement of nodulation by benzyladenine was recently reported by Yahalom *et al.* (1990).

In preliminary experiments, alfalfa and vetch seedlings treated with 1 µg/L solutions of some cytokinins developed thick, short, hairy roots (Byl and Taller, unpublished data). Puppo and Rigaud (1978) reported a similar response by French-bean roots to *R. phaseoli* or exogenous cytokinins. This syndrome, dubbed the Tsr response is produced by sterile rhizobial supernatants on homologous host plant roots and requires *nod* gene induction (Zaat *et al.*, 1987). Tsr is also induced by the recently discovered plant-specific lipo-oligosaccharide signals which appear to be the bacterial "nodule organogenesis inducing principle" (Spaink *et al.*, 1991).

There is evidence that cytokinins are produced in root nodules. Nodules of legumes and nonlegumes have been reported to contain higher amounts of cytokinin than the root itself (Henson and Wheeler,1976, 1977; Syono *et al.*, 1976). However, other studies have found no difference between inoculated and uninoculated roots (Wang *et al.*, 1982). Badenoch-Jones *et al.* (1984) found no major difference in the metabolism of exogenously applied cytokinins in roots or effective and ineffective nodules. However, this does not rule out possible differences in endogenous cytokinin content. Letham *et al.* (1990) reported that cytokinin levels in the xylem sap of nodulated pigeonpea plants were twice that of uninoculated plants. Plants inoculated with the cytokinin overproducing strain of *Rhizobium* had 8 times more cytokinin riboside than normal nodulated plants. These data indicate that rhizobia produce cytokinins both in the free-living and symbiotic state, and can export cytokinin via the xylem. Thus, nodulation may affect plant growth by modifying the cytokinin economy of the plant in addition to supplying fixed nitrogen.

In a simple experiment performed some years ago, Puppo and Rigaud (1978) examined cytokinin production by *Rhizobium phaseoli* and *Phaseolus vulgaris* roots. Lacking technical advances in cytokinin analysis which have occurred in recent years, they found no cytokinin in the hydroponic culture solution from the roots nor in the bacterial culture filtrate. However, when roots and rhizobia were cultured together, cytokinin levels equivalent to 10 µg/L of kinetin were found after 3 days of coincubation. Given the limited characterization of the cytokinins found, the source of the cytokinin activity cannot be determined. This symbiosis-induced cytokinin production may have been due to plant induction of bacterial cytokinin synthesis or to bacterial stimulation of plant cytokinin biosynthesis. Either of these possibilities is intriguing and deserving of further investigation.

Concluding remarks

All rhizobial strains examined in this laboratory produced cytokinins when grown on defined medium. Thus the question that remains is not *whether* rhizobia produce cytokinins but *why*. As discussed above, there is considerable circumstantial evidence for the involvement of cytokinins in legume nodulation. Elucidation of plant-specific lipo-oligosaccharide signals produced by rhizobia (Lerouge *et al.*, 1990a) make it unlikely that cytokinins act as the long-sought "nodule inducing principle". The mechanism of action of this nodule initiating signal is not yet known. It is possible that the lipo-oligosaccharide may act directly to cause root hair curling, cortical cell division and early nodulin induction. However, given the diversity of responses to this signal, it is also plausible that it may act by triggering a secondary signal (Long and Atkinson, 1990), perhaps leading to a change in the phytohormone balance of the root. Limited knowledge concerning the mechanism of action of phytohormones may hinder progress in understanding nodule initiation. However, elucidation of the mode of action of nodule inducing factors may provide information about signal transduction, hormone action and gene expression in plants.

The role of rhizobia-produced cytokinins remains unclear. Lipo-oligosaccharide-produced nodules exhibit anatomical and physiological features similar to bacterial induced nodules (Lerouge *et al.*, 1990b), suggesting that the oligosaccharide signal may be sufficient for nodulation. It has been proposed for *Agrobacterium* that the *tzs* gene, inducible by plant phenolics, may not be necessary for infection, but may affect the efficiency of transformation by influencing host range, susceptibility to infection or tumor initiation (Akiyoshi *et al.*, 1987; Zahn *et al.*, 1990). The effects of cytokinins on nodulation suggest that rhizobial cytokinins may have similar roles. Confirmation of a role for rhizobial cytokinins in nodulation awaits availability of a strain devoid of these compounds.

Acknowledgement

Work in this laboratory was funded in part by the Feinstone Microbiology Research Fund at MSU and by The Martin Himmel Health Foundation.

References

Akiyoshi, D.E., Regier, D.A. & Gordon, M.P. (1987) *J. Bacteriol.* **169**, 4242-4248.

Allen, E.K., Allen, O.N. & Newman, A.S. (1953) *Am. J. Bot.* **40**, 429-435.

Arora, N., Skoog, F. & Allen, O.N. (1959) *Am. J. Bot.* **46**, 610-613.

Badenoch-Jones, J., Rolfe, B.G. & Letham, D.S. (1984) *Plant Physiol.* **74**, 239-246.

Bauer, W.D., Bhuvaneswari, T., Calvert, H.E., Law, I.J., Malik, N.S.A., & Vesper, S.J., (1985) In: Nitrogen Fixation Research Progress, Evans, H.J. Bottomly, P.J. & Newton, W.E. (eds), Martinus Nijhoff Publishers, Dordrecht, The Netherlands pp 247-253.

Cooper, J.B. & Long, S.R. (1988) In: Physiology and Biochemistry of Plant Microbial Interactions, Keen, N.T., Kosuge, T. & Wallings, L.L. (eds), American Society of Plant Physiologists, Rockville, Maryland p 148.

de Bruijn, F.J. (1989) In: Plant Microbe Interactions, v. 3, Kosuge, T. & Nester, E.W. (eds), McGraw-Hill, New York p 457-504.

Greene, E.M. (1980) *Bot. Rev.* **46**, 25-74.

Henson, I.E. & Wheeler, C.T. (1976) *New Phytol.* **76**, 433-439.

Henson, I.E. & Wheeler, C.T. (1977) *J. Exp. Bot.* **28**, 1076-1086.

Hirsch, A.M., Bhuvaneswari, T.V., Torrey, J.G. & Bisseling, T. (1989) *Proc. Natl. Sci. USA* **86**, 1244-1248.

Lerouge, P., Roche, P., Faucher, C., Maillet, F., Truchet, G., Prome, J.C. & Denairie, J. (1990a) *Nature* **344**, 781-784.

Lerouge, P., Roche, P., Prome, J.C., Faucher, C., Vasse, J., Maillet, F., Camut, S., de Billy, F., Barker, D.G., Denarie, J. & Truchet, G. (1990b) In: Nitrogen Fixation: Achievements and Objectives, Gresshoff, P.M., Roth, L.E., Stacey, G. & Newton, W.E. (eds), Chapman and Hall, New York pp 177-186.

Letham, D.S., Parker, C.W., Zhang, R., Singh, S., Upadhyaya, M.N., Dart, P.J. & Palni, L.M.S. (1990) In: Plant Growth Substances 1988, Pharis, R.P. & Rood, S.B. (eds), Springer-Verlag, Berlin pp 275-281.

Long, S.R. & Cooper, J. (1988) In: Molecular Genetics of Plant-Microbe Interactions, Palacios, R. & Verma, D.P.S. (eds), APS Press, St. Paul, Minnesota pp 163-178.

Long, S.R. & Atkinson, E.M. (1990) *Nature* **344**, 712-713.

Medford, J.I., Horgan, R., El-Sawi, Z. & Klee, H.J. (1989) *Plant Cell* **1**, 402-413.

Morris, R.O., (1986) *Annu. Rev. Plant Physiol.* **37**, 509-538.

Nirunsuksiri, W. & Sengupta-Gopalan, C. (1988) In: Physiology and Biochemistry of Plant-Microbial Interactions, Keen, N.T., Kosuge, T. & Wallings, L.L. (eds) American Society of Plant Physiologists, Rockville, Maryland pp 171-172.

Pain, A.N. (1979) *J. Appl. Bacteriol.* **47**, 53-64.

Powell, G.K., Hommes, N.G., Kuo, J. Castle, L.A. & Morris, R.O. (1988) *Mol. Plant-Microbe Interact.* **1**, 235-242.

Puppo, A. & Rigaud, J. (1978) *Physiol. Plant.* **42**, 202-206.

Rodriguez-Barrueco, C. & Bermudez de Castro, F. (1973) *Physiol. Plant.* **29**, 277-280.

Spaink, H.P., Geiger, O., Sheeley, D.M., van Brussel, A.A.N., York, W.S., Reinhold, V.N., Lugtenberg, B.J.J. & Kennedy, E.P. (1991) In: Advances in Molecular Genetics of Plant-Microbe Interactions, Hennecke, H. & Verma, D.P.S. (eds), Kluwer Academic Publishers, Dordrecht, The Netherlands pp 142-149.

Sturtevant, D.B. & Taller, B.J. (1989) *Plant Physiol.* **89**, 1247-1252.

Syono, K., Newcomb, W. & Torrey, J.G. (1976) *Can. J. Bot.* **54**, 2155-2162.

Taller, B.J. & Sturtevant, D.B. (1991) In: Advances in Molecular Genetics of Plant-Microbe Interactions, Hennecke, H. & Verma, D.P.S. (eds), Kluwer Academic Publishers, Dordrecht, The Netherlands pp 215-221.

Wang, T.L., Wood, E.A. & Brewin, N.J. (1982) *Planta* **155**, 350-355.

Yahalom, E., Okon, Y. & Dovrat, A. (1990) *Can. J. Microbiol.* **36**, 10-14.

Zaat, S.A.J., Van Brussel, A.A.N., Tak, T., Pees, E. & Lugtenberg, B.J.J. (1987) *J. Bacteriol.* **169**, 3388-3391.

Zahn, X., Jones, D.A. & Kerr, A. (1990) *Plant Mol. Biol.* **14**, 785-792.

Nodulation in the absence of *Rhizobium*

Gustavo Caetano-Anollés, Priyavadan A. Joshi and
Peter M. Gresshoff

*Plant Molecular Genetics, Institute of Agriculture and Center for Legume
Research, The University of Tennessee, Knoxville, TN 37901-1071, USA*

Introduction

A wide variety of legumes are able to establish symbiotic association with nitrogen-fixing *Rhizobium* and *Bradyrhizobium* bacterial species. Through a complex series of interactions these soil bacteria infect the roots of the host plant and induce the formation of a specialized organ, the nodule (Caetano-Anollés and Gresshoff, 1991). To initiate this process the host root secretes substances that act as chemoattractants and inducers of gene expression in the bacteria. In turn, the bacteria move towards and associate with the host root surface, secrete factors that alter the pattern of the original plant inducers and release nodulation signals that elicit the first two visible morphological responses in the host, root hair curling and cortical cell division. The bacteria are then entrapped between cell wall surfaces, usually by the pronounced curling of a root hair, and penetrate through the underlying host cell wall to produce a tubular structure called the infection thread. The continuous deposition of cell wall-like material around the bacterial cells and their multiplication results in elongation and branching of the infection thread towards the dividing cortical root cells. In the meantime, the dividing cortical cells give rise to a meristematic center beneath the infection site, the nodule primordium. The bacteria are finally released into dividing host cells and packaged within host plasmamembrane where they differentiate into bacteroids and fix atmospheric nitrogen into ammonia in exchange for fixed carbon and other nutrients.

This complex process is carefully regulated at different stages during infection, as judged by the number of bacterial and plant mutants that have been isolated which differentially arrest nodule development. Such mutants can dissect nodulation in its different regulatory components and provide insight into the role of bacterial and plant signals that are crucial for the establishment of a functional nodule.

Bacteria-free organized nodular structures

Rhizobium meliloti is able to induce cellular redifferentiation in the root cortex of alfalfa leading to cell division and nodule organogenesis (Truchet et al, 1980; 1984; Hirsch et al, 1984; Dudley et al, 1987; Caetano-Anollés and Gresshoff, 1990). The inner root cortical cells divide anticlinally and then periclinally and initiate a nodule primordium that finally forms a persistent nodule meristem in its distal end (Dudley et al, 1987). The meristem gives rise to a central tissue which contains both infected and uninfected cells. A number of agents can produce nodules without infection. For example, ineffective bacteria-free nodules are produced in response to *R. meliloti* mutants defective in exopolysaccharide synthesis (*exo* mutants) (Finan et al, 1985; Leigh et al, 1987) and *Agrobacterium tumefaciens* transconjugants carrying *R. meliloti* nodulation genes (Wong et al, 1983; Truchet et al, 1984; Hirsch et al, 1984; 1985). These nodules resemble nitrogen-fixing nodules in that a central tissue is surrounded by nodule parenchyma, an endodermis and a nodule cortex, and contains peripheral vascular bundles. Nodulins that are associated with nodule morphogenesis, like the hydroxyproline-rich glycoprotein Enod2, express in the nodular tissue of these bacteria-free nodules (Dickstein et al, 1988; Norris et al, 1988; Van de Wiel et al, 1990).

Root-derived structures that fulfill most of the histological and molecular criteria defining an indeterminate nodule can be also formed in the absence of *Rhizobium*. These structures appear in alfalfa roots by treatment with auxin-transport inhibitors such as *N*-(1-naphtyl) phtalamic acid (NPA) and 2,3,5-triiodobenzoic acid (TIBA) (Hirsch et al, 1989), when *Rhizobium* is separated from the roots by filter membranes (Kapp et al, 1990), or spontaneously (Truchet et al, 1989). *Rhizobium* and *Bradyrhizobium* secrete cytokinins (Philips and Torrey, 1971; Sturtevant and Taller, 1989; Taller and Sturtevant, 1991) which are able to induce subepidermal cell division in soybean (Bauer et al, 1985). *Rhizobium* can also produce extracellular nodulation signals that induce cortical cell division (Hollingsworth et al, 1990) and empty nodules (Lerouge et al, 1990a,b), or low-molecular-weight substances that stimulate mitosis in plant protoplasts (Schmidt et al, 1988).

Nodulation in the absence of *Rhizobium* (Nar)

It was not until recently that the existence of 'genuine' nodules formed spontaneously in the absence of *Rhizobium* was recognized (Truchet et al, 1988). Cytological studies showed that these nodules were organized structures rather than abnormal lateral root outgrowths that retained the normal histological features of a normal indeterminate alfalfa nodule (Truchet et al. 1989; Caetano-Anollés et al, 1991; Joshi et al, 1991). Spontaneous nodules were single-to-multilobed structures with no inter- or intracellular bacteria, composed of nodule meristems, cortex, endodermis and central zone with vascular strands.

While no viable bacteria were recovered from surface sterilized nodules the content of crushed non-sterilized nodules was unable to induce nodulation, indicating that bacteria were not the etiological agent of spontaneous nodulation.

Only a small subpopulation (1-6%) of the several different alfalfa cultivars studied so far developed nodules in the absence of *Rhizobium* and combined nitrogen (Truchet et al. 1989; Caetano-Anollés et al, 1991). Spontaneous nodules formed mainly on the primary root of alfalfa cv. Vernal, as early as 10 d after seed imbibition, in the region susceptible to *Rhizobium* infection between 4 and 6 d post-germination (Caetano-Anollés et al, 1991). Nodules emerged with a rate comparable to *Rhizobium*-induced nodules, but a week later than first nodules formed when 5-d-old plants were inoculated. While spontaneous nodules did not support plant growth or fix atmospheric nitrogen, plants grown in 15 mM nitrate did not form visible nodules (Truchet et al. 1989; Caetano-Anollés et al, 1991). As with bacteria-free nodules induced by *R. meliloti exo* mutants, Enod2 was transcribed in spontaneous nodules (Truchet et al, 1989). This morphogenetic marker adds evidence to suggest that the nodule-like structures formed in the absence of *Rhizobium* are indeed genuine nodules.

Spontaneous nodules elicit autoregulatory responses

The formation of nodules in alfalfa is controlled by a systemic feedback regulatory (or autoregulatory) mechanism that suppresses nodulation in younger parts of the root system (Caetano-Anollés and Bauer, 1988). Nodule excision experiments demonstrated that feedback suppression was exerted at the level of nodule initiation rather than during infection development (Caetano-Anollés and Gresshoff, 1991). Bacteria-free nodular structures induced by *R. meliloti exo* mutants were also able to elicit and be the target of this systemic response (Caetano-Anollés et al, 1990), indicating that extensive infection was not required for the elicitation of autoregulation.

To examine whether autoregulatory responses are induced by diffusible bacterial substances or result indirectly from the formation of a nodule primordium, we studied the ability of nodules formed in the absence of *Rhizobium* to control *Rhizobium*-induced nodulation. The formation of spontaneous nodules on the primary root suppressed nodulation on lateral roots after inoculation with *R. meliloti* (Caetano-Anollés et al, 1991). Excision of spontaneous nodules at inoculation eliminated the suppressive response. The suppression of *Rhizobium* nodulation by the spontaneous nodules was systemic and rapid, as judged by the ability of spontaneous nodules on primary roots to suppress nodulation and the distribution of *Rhizobium*-induced nodules on lateral roots (Fig. 1). While the presence of spontaneous nodules on the primary root allowed for almost the same number of nodules to be formed on each lateral root, the fraction of lateral roots that were nodulated decreased significantly ($P<0.05$). This effect was not related to a decrease in the number of lateral roots.

Our results indicate that *Rhizobium* is not required for elicitation of feedback suppression of nodulation in the host. Excision of spontaneous nodules eliminated the suppressive response indicating that plants with spontaneous nodules were not intrinsically defective in nodulation.

We were unable to find white single-to-multilobed spontaneous nodules forming in plants that were inoculated with wild-type *R. meliloti* early during plant development. Inoculated plants formed only pink nitrogen-fixing indeterminate nodules. This suggests that spontaneous nodules are subject to active suppression by *Rhizobium*-induced nodules appearing shortly after inoculation.

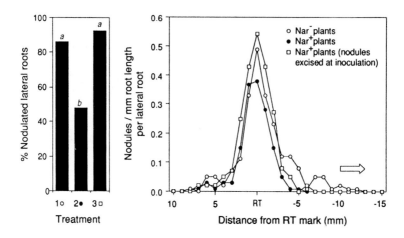

Figure 1. Suppression of Rhizobium nodulation by spontaneous nodules. Sets of 15-26 alfalfa (M. sativa cv. Vernal) plants with spontaneous nodules, and 75 nonnodulated plants were inoculated with 10^6 bacteria·plant^{-1} and examined for the appearance of nodules at regular intervals. At the time of inoculation the position of each lateral root tip (RT) was marked on the plastic surface of the pouch, and spontaneous nodules in one of the treatments were excised and removed from the pouch. The location of each individual nodule appearing on lateral roots was determined relative to the RT mark, 22 d after inoculation. Nodules appearing in 61-104 lateral roots from 9-23 alfalfa plants were examined. The rate of lateral root elongation was 3.2 ± 0.5 mm·d^{-1}, and the average number of lateral roots 10.1 ± 1.5 lateral roots·plant^{-1}. Plants with spontaneous nodules (Nar$^+$) had an average of 4.5 ± 0.2 nodules·plant^{-1}. The direction of root growth is shown by an arrow. Bars headed by the same letter are not significantly different (p=0.05).

Ontogeny and ultrastructure of spontaneous nodules

It is important to assess correctly the definition of the nodules formed spontaneously in alfalfa, specially in view of the recent reports on rice, wheat and rape nodulation (Al-Mallah et al, 1989; Cocking et al, 1990) and the discovery that extracellular bacterial nodulation signals induce cell divisions and nodule-like structures in roots (Hollinsworth et al, 1990; Lerouge et al, 1990a,b). The ontogeny, morphogenesis and structural organization of the organ rather than its terminal function can provide the suitable criterion.

Both the histological organization and the ontogeny of spontaneous nodules resembled that of normal indeterminate alfalfa nodules (Joshi et al, 1991). Anticlinal followed by periclinal divisions of redifferentiated cells in the inner root cortex resulted in the formation of a nodular meristem. Repeated cell division produced a group of compactly arranged daughter cells with prominent nucleus and nucleolus, high cytoplasmic density, and abundant starch grains as compared to elongated adjacent cortical cells (Fig. 2). Finally, a dome-shaped nodular meristem with highly dense cells of uneven size was formed and protruded outwards through the surrounding cortical tissue. Cell divisions in the meristem formed large and small cells which persisted in the central zone. Large cells had a large central vacuole and cytoplasm packed with starch. Few starch grain bodies were found in the small cells but instead they harbored abundant lipid bodies. While infected cells in a normal nodule contain a continuous network of rough-endoplasmic reticulum, this membrane system was sparse or discontinuous in spontaneous nodules. These results indicate that alfalfa plants have the endogenous ability to induce cell division in the root cortex and develop bacteria-free nodules with distinct cell types.

The preferential deposition of starch into large cells in the absence of *Rhizobium* suggests that the conversion of translocated sugars into starch and their storage in differentially expanded cells are inherent parts of nodule growth ontogeny rather than part of a developmental cascade triggered by *Rhizobium*. It also suggests that the ancestral nodule may have initially functioned as a carbon-storage organ that in time evolved to include the nitrogen-fixing bacteria. It is then feasible to find similar spontaneous nodule-like structures in other legume and non-legume plant species, perhaps with a carbon-storage function.

While spontaneous nodules are elongate and usually multilobed structures as those induced by wild-type *R. meliloti*, the majority of nodules induced by *R. meliloti exo* mutants are small and broad and appear like beads on a string. Meristematic activity in spontaneous nodules is highly confined (Joshi et al, 1991). On the other hand, nodules induced by *exo* mutants have their meristems spread along the distal end of the nodule with little peripheral and central tissue formed perpendicularly to the root axis (Van de Wiel et al, 1990). This gives their characteristic appearance. Despite the many morphological and molecular similarities, it is important to note that these two bacteria-free nodular structures cannot be considered homologous.

Nar and bacterial nodulation signals

Although little is known about the mechanisms that determine and control nodule initiation, there is evidence that nodulation (*nod*)gene related extracellular signaling is required to elicit the early responses of root hair deformation and curling in alfalfa (Faucher et al. 1989; Banfalvi and Kondorosi, 1989). One of such signals, a sulphated and acylated lipo-oligosaccharide, can also induce cortical cell division and nodulation (Lerouge et al, 1990a,b). This alfalfa-specific nodulation signal, NodRm-1, requires the *nodABC* genes for its synthesis and the host-range *nodH* and *nodQ* genes for its activity. The nodP and nodQ gene products have sulphurylase activity (Schwedock and Long, 1990) and with the nodH gene product could activate inorganic sulphate and introduce it into a precursor of NodRm-1, turning it from vetch-specific into alfalfa-specific.

To determine if nodule initiation was limited by the concentration and/or quality of certain bacterial nodulation signal substances, *R. meliloti nod* gene mutants were co-inoculated with suboptimal concentrations of wild-type bacteria (Caetano-Anollés and Bauer, 1990). Pairs of host-specificity mutants help wild-type *R. meliloti* initiate nodule formation, apparently by providing a sustained supply of different but complementary extracellular signals required for the induction of appropriate host responses during nodule initiation. These signals act synergistically to favor nodule initiation.

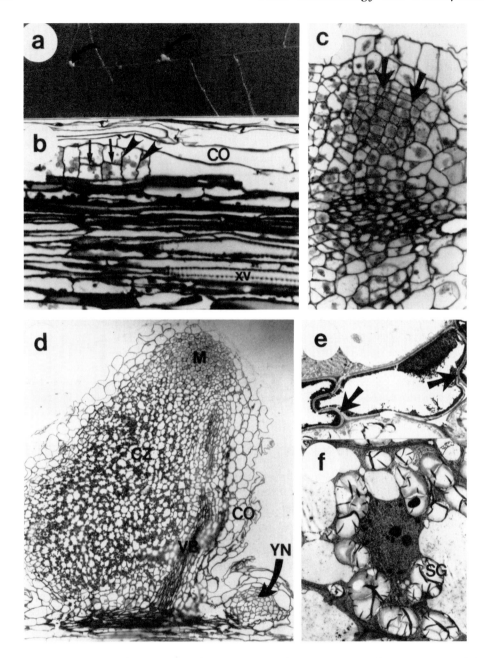

Figure 2. Nodulation in the absence of Rhizobium. a. Alfalfa primary root with spontaneous nodules (arrows). b. Longitudinal root section showing anticlinal (arrowheads) and periclinal (arrows) cell division of redifferentiated cells in the inner cortex; x440. c. Root cross-section showing repeated division of cortical cells (arrows) producing a young nodule meristem that pushes the root cortex outward; x280. d. Longitudinal view of a mature spontaneous nodule and a young emerging nodule; x70. e. Endodermal layer showing Casparian strips on the side walls (arrows); x2250. f. Central zone cells with starch grain bodies and no bacteria; x2000. Abbreviations: CO, cortex; CZ, central zone; M, meristem; SG, starch grain; YN, young nodule; VB, vascular bundle; xv, xylem vessel.

The plurality of signals was further investigated. Inoculation of alfalfa plants with *R. meliloti nodH* nonnodulating mutants resulted in a 5-fold increase in the number of plants with empty nodules when compared to sham inoculations (Caetano-Anollés and Gresshoff, in preparation). Inoculation with *nodABC* mutants gave no such increase. These results suggest that other bacterial signals, different from but perhaps precursors of NodRm-1, are able to induce host responses in alfalfa. That extracellular signals determine the efficiency of Nar induction further suggests co-evolution of plant and bacteria to optimize the development of the symbiosis.

Inheritance of Nar

Nodulation in the absence of *Rhizobium* was maintained during clonal propagation of the alfalfa plants (Truchet et al, 1989). However, the nodulation pattern in cultivar Gemini was variable between individuals, with some plants nodulating profusely and other forming only few multilobed nodules after a long period of cultivation. This suggests that one or more genetic determinants can condition the Nar phenotype.

Nodulation in the absence of *Rhizobium* and the formation of empty nodules upon inoculation with nonnodulating bacterial mutants segregated in progeny plants of Vernal alfalfa (Caetano-Anollés and Gresshoff, in preparation). Almost all F_1 plants from crosses between Nar+ individuals or plants with empty nodules induced by *Rhizobium* mutants, and almost all S_1 progeny formed empty nodules resembling genuine spontaneous nodules. The number and location of these nodules within the root system remained the same, possibly due to a strong genetic control over the ability of an alfalfa plant to control spontaneous nodulation. The mode of inheritance indicated that a single dominant genetic element controls Nar in this tetraploid alfalfa.

Concluding remarks

Different criteria define the nodule-like structures formed in the absence of *Rhizobium* as genuine nodules (Table 1). Both spontaneous and normal nodules share the same histological organization and ontogeny, expressed nodulins associated with nodule morphogenesis, induced feedback suppressive responses, and were inhibited by nitrate. Our results suggest that the presence of *Rhizobium* is not required for nodule organogenesis and the elicitation of nodulation control. Since Nar is a heritable character, probably dominant in nature, some genotypes are able to express the "nodulation cascade" spontaneously whereas other genotypes depend on extracellular nodulation signals. Our observations suggest that bacteria and their *nod* genes do not condition the ontogeny and morphogenesis of spontaneous nodulation.

Most features of a nodule are not under the inductive control of *Rhizobium* but represent an internal developmental program, whose expression is presumably optimized in some homozygous recessive plant genotypes by *Rhizobium*.

Spontaneous nodulation has not yet been reported in other legumes. Is the heterozygous nature of the the alfalfa population responsible for the maintenance of the Nar phenotype in the seed population? If so, would some individuals from an heterozygous clover or *Glycine* population show spontaneous nodules? Is the stored starch available as a carbon source? Are there homologous proteins in the membrane enclosing the starch grain and the peribacteroid membrane? Do Nar+ and Nar- plants produce different phytohormones? Are there detectable biochemical differences in the presumed signal-response chain? We are now attempting to address some of these questions, to find a plant that is amenable for genetic studies, and to determine if Nar is a widespread phenomenon in nature.

Table 1. Definition of the Nar phenotype

Origin:	*Inner cortex near pericycle*
Structure:	*Cortex, endodermis, central tissue with large and small cells, vascular system*
Function:	*Starch storage; enod2 expression*
Regulation:	*Nitrate inhibition*
	Autoregulation
Stability:	*Vegetative*
	Sexual

While Nar can be the product of a yet undefined genetic element present in certain alfalfa genotypes (a non-culturable endophytic bacterium, a mycoplasm, or a plant virus or viroid), it appears most likely that it is caused by an alfalfa gene which is 'constitutive' for one of the steps in the signal transduction chain starting with the *Rhizobium* derived signals and ending in nodule formation. Such genotypes are reminiscent of disease mimics in corn (*Zea mays*), which exhibit disease symptoms (stripping of leaves, etc.) in the absence of the normally associated pathogen. The question arises about the stability of such a genetic element in a population. Spontaneous nodules are clearly unable to fix nitrogen and hence should be viewed as a useless appendage. Presumably, continued selection for rapid growth and yield during domestication should have eliminated a gene that confers its phenotype even in an heterozygous condition in a tetraploid genetic background. An understanding of such evolutionary aspects will help with the clarification of the apparent species distribution of nodulation among angiosperms.

Acknowledgement

This work was supported by an endowment to the Racheff Chair of Excellence of the University of Tennessee, and in part by a grant from Hi-Bred Pioneer International and the Soybean Promotion Board, Haskinville, Tenn., USA. We thank Dr. Effin T. Graham for helpful discussions.

References

Al-Mallah, M.K., Davey, M.R. & Cocking, E.C. (1989) *J. Exp. Bot.* **40**, 473-478.

Banfalvi, Z. & Kondorosi, A. 1989. *Plant Mol. Biol.* **13**, 1-12.

Bauer, W.D., Bhuvaneswari, T.V., Calvert, H.E., Law, I.J., Malik, N.S.A. & Vesper, S.J. (1985) In: Nitrogen fixation research progress. Evans, H.J., Bottomley, P.J., Newton, W.E. (ed.), Nijhoff, Dordretch, pp. 247-253.

Caetano-Anollés, G. & Bauer, W.D. (1988) *Planta* **175**, 546-557.

Caetano-Anollés, G. & Bauer, W.D. (1990) *Planta* **181**, 109-116.

Caetano-Anollés, G.& Gresshoff, P.M. (1991) *Plant Physiol.* **95**, 366-373.

Caetano-Anollés, G.& Gresshoff, P.M. (1991) *Annu. Rev. Microbiol.*, **45**, 345-382.

Caetano-Anollés, G., Joshi, P.A. & Gresshoff, P.M. (1991) *Planta* **183**, 77-82.

Caetano-Anollés, G., Lagares, A. & Bauer, W.D. (1990) *Plant Physiol.* **92**, 368-374.

Cocking, E.C., Al-Mallah, M.K., Benson, E. & Davey, M.R. (1990) In: Nitrogen fixation: achievements and objectives. Gresshoff, P.M, Roth, E., Stacey, G., Newton, W.E. (ed.), Chapman & Hall, New York, pp. 813-823.

Dickstein, R., Bisseling, T., Reinhold, V.N. & Ausubel, F.M. (1988) *Genes Dev.* **2**, 677-687.

Dudley, M.E., Jacobs, T.W. & Long, S.R. (1987) *Planta* **171**, 289-301.

Faucher, C., Camut, S., Dénarié, J. & Truchet, G. (1989) *Mol. Plant-Microbe Interact.* **2**, 291-300.

Finan, T.M., Hirsch, A.M., Leigh, J.A., Johansen, E., Kuldau, G.A., Deegan, S., Walker, G.C. & Signer, E.R. (1985) *Cell* **40**, 869-877.

Hirsch, A.M., Wilson, K.J., Jones, J.D.G., Bang, M., Walker, V.V. & F.M. Ausubel (1984) *J. Bacteriol.* **158**, 1133-1143.

Hirsch, A.M., Drake, D., Jacobs, T.W. & Long, S.R. (1985) *J. Bacteriol.* **161**, 223-230.

Hirsch, A.M., Bhuvaneswari, T.V., Torrey, J.G. & Bisseling, T. (1989) *Proc. Natl. Acad. Sci. USA* **86**, 1244-1248.

Hollingsworth, R.I., Philip-Hollingsworth, S. & Dazzo, F.B. (1990) In: Nitrogen fixation: achievements and objectives. Gresshoff, P.M, Roth, E., Stacey, G., Newton, W.E. (ed.), Chapman & Hall, New York, pp. 193-198.

Joshi, P.A., Caetano-Anollés, G., Graham, E.T. & Gresshoff, P.M. (1991) *Protoplasma* **162**, 1-11.

Kapp, D., Niehaus, K., Quandt, J., Müller, P. & Pühler, A. (1990) *Plant Cell* **2**, 139-151.

Leigh, J.A., Reed, J.W., Hanks, J.F., Hirsch, A.M. & Walker, G.C. (1987) *Cell* **51**, 579-587.

Lerouge, P., Roche, P., Faucher, C., Maillet, F., Truchet, G., Promé, J. & Dénarié, J. (1990a) *Nature* **244**, 781-784.

Lerouge, P., Roche, P., Promé, J., Faucher, C., Vasse, J., Maillet, F., Camut, S., de Billy, F., Barker, D.G., Dénarié, J. & Truchet, G. (1990b) In: Nitrogen fixation: achievements and objectives. Gresshoff, P.M, Roth, E., Stacey, G., Newton, W.E. (ed.), Chapman & Hall, New York, pp. 177-186.

Norris, J.H., Macol, L.A. & Hirsch, A.M. (1988) *Plant Physiol.* **88**, 321-328.

Phillips, D.A. & Torrey, J.G. (1971) *Plant Physiol.* **49**, 11-15.

Schmidt, J., Wingender, R., John, M., Wieneke, U. & Schell, J. (1988) *Proc. Natl. Acad. Sci. USA* **85**, 8578-8582.

Schwedock, J. & Long, S.R. (1990) *Nature* **348**, 644-647.

Sturtevant, D.B. & Taller, B.J. (1989) *Plant Physiol.* **89,** 1247-1252.

Taller, B.J. & Sturtevant, D.B. (1991) In: Biochemical and genetic analysis of gene expression in plants and bacteria. Gresshoff, P.M. (ed.), Chapman & Hall, New York, (this volume).

Truchet, G., Michel, M. & Dénarié, J. (1980) *Differentiation* **16,** 163-172

Truchet, G., Rosenberg, C., Vasse, J., Julliot, J.S., Camut, S. & Dénarié, J. (1984) *J. Bacteriol.* **157,** 134-142.

Truchet, G., Vasse, J., Odorico, R., de Billy, F., Camut, S. & Huguet, T. (1988) In: Molecular genetics of plant-microbe interactions. Palacios, R., Verma, D.P.S., APS Press, Minnesota, pp. 179-180.

Truchet, G., Barker, D.G., Camut, S., de Billy, F., Vasse, J. & Huguet, T. (1989) *Mol. Gen. Genet.* **219,** 65-68.

Van de Wiel, C., Norris, J.H., Bochenek, B., Dickstein, R., Bisseling, T. & Hirsch, A.M. (1990) *Plant Cell* **2,** 1009-1017.

Wong, C.H., Pankhurst, C.E., Kondorosi, A. & Broughton, W.J. (1983) *J. Cell Biol.* **97,** 787-794.

Altered tryptophan biosynthesis in *Bradyrhizobium japonicum* gives enhanced nodulation and nitrogen fixation

L. David Kuykendall and William J. Hunter

Agricultural Research Service, United States Department of Agriculture, Beltsville, Maryland 20705 and Fort Collins, CO 80522, USA

Introduction

A specific *Bradyrhizobium japonicum* tryptophan auxotroph was discovered to produce prototrophic revertants with enhanced symbiotic capabilities. This finding constituted a new method of increasing symbiotic nodulation and nitrogen fixation in soybean (Kuykendall and Hunter, U.S. Patent Pending #07/325,184).

Several investigators have isolated mutants of *B. japonicum* with altered tryptophan metabolism and have found that such mutants often have altered symbiotic properties. Usually, it has been a loss of nodulation and nitrogen fixation ability that was observed. *B. japonicum* I-110ARS mutants that are resistant to 5-methyltryptophan have been isolated and most, though not all, such mutants were poor nodulators and nitrogen fixers (Hunter, 1987 and 1990). The mechanism of resistance to 5-methyltryptophan is unknown in *B. japonicum*. However, in *Escherichia coli* and *Bacillus subtilis* this toxic tryptophan analog often selects for cells that constitutively overproduce tryptophan pathway products (Hoch et al, 1971; Shiio et al, 1972). Tryptophan catabolic variants of *Bradyrhizobium* sp. (soybean) strain L259 can have improved symbiotic properties (Kaneshiro and Kwolek, 1985) under specific conditions (Kaneshiro and Nicholson, 1989). Characteristically, these catabolic mutants, when grown in the presence of tryptophan, degrade tryptophan rapidly and accumulate large amounts of indole compounds. It is thought that this increased production of indole compounds alters the symbiotic properties of these bacteria (Kaneshiro et al, 1983; Kaneshiro and Kwolek, 1985).

Tryptophan auxotrophs of *B. japonicum* I-110ARS have also been studied. Characterization studies (Wells and Kuykendall, 1983) have placed these auxotrophs into biochemical classes based on the defects within the tryptophan biosynthetic pathway of the mutant (Figure 1). Mutants with enzyme defects in the tryptophan pathway before tryptophan synthase were not capable of developing normal appearing nodules. However, mutants that were defective in tryptophan synthase, the last enzyme of the pathway, were able to form nodules that appeared normal.

The nodules formed by this latter group, while normal in appearance, were Fix⁻ (Kummer and Kuykendall, 1989). This evidence indicates that the ability to make indole-3-glycerol phosphate, the product of indole glycerol phosphate synthase, the third enzyme of the pathway, may be necessary for nodule development (Kummer and Kuykendall, 1989).

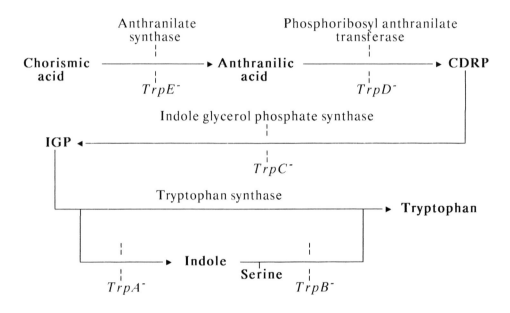

Figure 1. The tryptophan biosynthetic pathway. Products and substrates are depicted in bold type and enzymes in normal type with their corresponding genes in italics.

While tryptophan auxotrophs of *B. japonicum* I-110ARS had defective symbiotic properties, a prototrophic revertant of a tryptophan auxotroph was obtained that showed a symbiotic relationship with soybean that was superior to that formed by the wildtype strain, *B. japonicum* I-110ARS (Hunter and Kuykendall, 1990). Strain TA-11, the parent of the new superior strain, was incapable of forming bona fide nodules as an auxotroph. However, when sufficient numbers of bacterial cells of strain TA-11 were applied as inoculum to soybean, or when selection was made in vitro on minimal medium, prototrophic revertants capable of forming effective nodules were obtained. Revertantcontaining nodules were discovered to provide a particularly suitable means of selecting and isolating prototrophic revertants having an enhanced ability to fix more nitrogen (Kuykendall and Hunter, U.S. Patent Pending #07/325,184).

Results

Properties of tryptophan auxotrophs

The parent strain used in these studies, *B. japonicum* strain I-110ARS, carried antibiotic resistance markers for azide, rifampin, and streptomycin, and was wildtype in symbiotic capability (Kuykendall and Weber, 1978). These resistance markers made it possible to easily distinguish I-110ARS mutants from contaminants. Nitrous acid mutagenesis was used to isolate eleven tryptophan requiring mutants from the parent *B. japonicum* strain (Kuykendall, 1987). Wells and Kuykendall (1983) used enzyme deficiencies and growth media supplementation requirements to characterize these eleven tryptophan auxotrophs. Mutants were found with defects in each of the four steps of the tryptophan pathway, and at least one mutant, TA-8, had several enzyme deficiencies (Table 1). We suspect that TA-8 is a regulatory mutant. Eight auxotrophic strains, those with defects at any of the first three pathway steps, did not form fully developed nodules on soybean as auxotrophs. Revertants did nodulate. Only the three mutants that were specifically defective in tryptophan synthase were able to form normal appearing nodules as auxotrophs.

Table 1. Biochemical and Symbiotic Properties of Bradyrhizobium japonicum Tryptophan Auxotrophs[a]

Strain	Enzyme deficiency[b]	Defective pathway step	Nodulation[c]
TA-1	TS	4	+
TA-2	TS	4	+
TA-3	AS	1	−
TA-4	AS	1	−
TA-5	PRT	2	−
TA-6	PRT, low IGPS	2	−
TA-7	AS	1	−
TA-8	AS, IGPS, TS	1,3,4	−
TA-9	AS	1	−
TA-10	TS	4	+
TA-11	IGPS	3	−

[a] *All, except strain TA7, are known from enzyme assays.*
[b] *TS = tryptophan synthase; AS = anthranilate synthase; PRT = phosphoribosyl transferase; IGPS = indole glycerol phosphate synthase*
[c] *Nodulation by the Trp mutants; = no nodules present that contained the auxotroph, but revertants to Trp+ do form nodules.*

The TA-11 strain was of special interest, as a revertant of this strain had improved symbiotic properties. The TA-11 strain is known to be deficient in a single enzyme of the tryptophan biosynthetic pathway, indole glycerol phosphate synthase. Due to this enzyme deficiency, this strain is unable to make indole3glycerol phosphate and is thus unable to make tryptophan (Wells and Kuykendall, 1983). Tryptophan supplementation is required for growth. Cells of this isolate form developmentally abortive root nodules as aminoacid requiring mutants (Kummer and Kuykendall, 1989). The TA-11 auxotroph has been deposited in the Agricultural Research Service Culture Collection (NRRL), Peoria, IL 61604 USA, under the designation NRRL B-18465.

Properties of the TA11NOD+ revertant

A number of prototrophic revertants of the TA-11 auxotroph have been isolated and details of studies on one, the TA-11NOD+ revertant, are given here. The TA-11NOD+ strain differs from its parent, the TA-11 Trp auxotroph, in that it tests positive for indole glycerol phosphate synthase activity (Hunter, unpublished), does not require tryptophan for growth and effectively nodulates soybean (Kummer and Kuykendall, 1989; Hunter and Kuykendall, 1990). Greenhouse studies have demonstrated that plants which received the TA-11NOD+ revertant had enhanced nitrogen fixation as evidenced by darker green color, higher dry weights, and larger nodule mass than did plants inoculated with wildtype bacteria (Hunter and Kuykendall, 1990). Subsequent experiments involving analysis of nitrogen accumulation further confirmed enhanced nitrogen fixation ability since they showed that, on a per plant basis, plants inoculated with TA-11NOD+ contained 52% more nitrogen, 34% more carbon and were 33% larger than plants inoculated with the wildtype progenitor strain I-110ARS. Plants treated with this mutant strain also had 56% more nodules and 41% more nodule mass than did plants inoculated with strain I-110ARS. Average nodule size and nitrogen fixed per gram of nodule were similar with either inocula. The improvement in nitrogen fixation observed with the TA-11NOD+ revertant correlated with an increase in nodule mass and an increased number of nodules per plant. Strain TA-11NOD+ has been deposited in the Agricultural Research Service Culture Collection (NRRL), Peoria, IL 61604 USA, under the designation NRRL B18466.

Methods for the selection of symbiotically superior mutants

For the growth of cells a HEPES (N-2-hydroxyethylpiperazine-N'-2-ethanesulfonic acid) and MES [2-(N-morpholino)ethanesulfonic acid] buffered nutrient solution (Cole and Elkan, 1973) was supplemented with 0.1% (w/v) L-arabinose to make minimal medium, and with arabinose and 1 g per liter yeast extract to make the A1E medium (Kuykendall, 1979). Arabinose was filter sterilized and added to the autoclaved base medium. For solid media, 1.5% (w/v) agar was added. Incubations were at 30 °C and broths were shaken at 130 rpm. Unless otherwise indicated, A1E medium was used for the growth and maintenance of all bacteria.

Selection of revertants using soybean plants was simple and reproducible. The following are details of an *in vivo* selection procedure useful for obtaining nodulating derivatives of strain TA-11. Plants were grown in vermiculite and provided with one-half strength nitrogen-free nutrient solution (Norris, 1964). Soybean seeds were surface-sterilized in 0.1% acidified HgCl 2 (Vincent, 1970). Leonard jars were used as containers to grow plants (Leonard, 1943). Five seeds per jar were each inoculated with 1.0 ml of late log phase culture of strain TA-11 and grown as described above. Plants were grown with a 16-hour day at 30 °C and an 8-hour night at 24 °C. Cotyledons were removed at eight days and seedlings were thinned to four per jar. After five weeks of growth, plant roots were excised at the first node. Normal-appearing nodules were carefully removed from the roots by hand and were surfacesterilized in freshly prepared 3% H_2O_2 solution for 1 hour. The nodules were then washed with sterile water to remove the disinfectant. Nodules were crushed with a sterile glass rod and the contents streaked onto both minimal and tryptophan-supplemented media. Prototrophic revertants derived from strain TA-11 were streaked onto selective media containing the appropriate antibiotics (500 μg/ml rifampicin and 1 mg/ml streptomycin) to confirm their genetic markers.

Selection of revertants on minimal media was as follows. Strain TA-11 was grown in A1E medium supplemented with tryptophan at 100 μg/ml for 4 days at 30 °C on a rotary shaker. Turbidity measurements, taken using a Klett-Summerson colorimeter (red filter), indicated a cell density of about 1.5×10^9 bacteria/ml by reference to a standard curve relating optical density in Klett units to cell count. Prototrophic revertants that are capable of forming nodules on soybean could be selected on minimal media. Mutants of strain TA-11 that have acquired independence from tryptophan as a requirement for growth occurred at a frequency of about one in 2×10^7 cells. Prototrophic revertants could be repeatedly selected and obtained from strain TA-11 (NRRL B-18465) using either the plant selection detailed above or this *in vitro* selection on minimal medium.

Isolation of the trpCD *genes from* B. japonicum

Our goal is to understand the molecular basis for alterations in tryptophan biosynthetic genes at the level of polynucleotide sequence. The *trpCD* genes of *B. japonicum* have been cloned for use in restriction mapping, sitedirected mutagenesis, and sequence comparison of mutant and wild-type genes. In order to isolate the relevant biosynthetic genes from *B. japonicum*, the *Rhizobium* meliloti cosmid pRmL104 was obtained from Graham Walker's lab at the Massachusetts Institute of Technology. The 4.7 kb *Eco*RI-*Pst*I fragment carrying *trpCD* was subcloned into pUC18 for use as a hybridization probe. Colony hyridization was performed on approximately 10^3 clones from the B. Chelm cosmid library of strain I-110 DNA (Adams, McClung and Chelm, 1984; Guerinot and Chelm, 1986). Two positive clones were identified as having the same 21 kb *Eco*RI fragment but in different orientations relative to the vector.

a b c d e f g

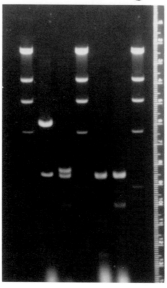

Figure 2. Restriction analysis of a recombinant plasmid carrying B. japonicum *DNA homologous to* trpCD *of* R. meliloti. *(a) lambda DNA digested with* HindIII, *(b) pDK1110 digested with* BamHI *and* EcoRI, *(c) pDK1110 digested with* BamHI, EcoRI, *and* HindIII, *(d) lambda DNA digested with* HindIII, *(e) pDK1110 digested with* BamHI, EcoRI, HindIII, *and* SalI, *(f) pDK1110 digested with* BamHI, EcoRI, HindIII, *and* SmaI, *(g) lambda DNA digested with* HindIII.

The homology was first delineated to a 4.7 kb *Eco*RI-*Bam*HI fragment. This fragment was subcloned into pUC18 (pDK1110). Restriction analysis of this recombinant plasmid (Figure 2) showed that the 4.7 kb *Eco*RI-*Bam*HI fragment was cleaved by *Hind*III to produce a 2.8 kb and a 1.9 kb fragment. The 1.9 kb fragment was cut by *Sal*I and the 2.8 kb fragment was cut twice by both *Sal*I and *Sma*I. This agarose gel was blotted and DNA hybridization was done using ^{32}P-labelled *R. meliloti trpCD* (Figure 3). The 2.8 kb *Bam*HI-*Hind*III fragment of pDK1110 carried the homology whereas the 1.9 kb *Eco*RI-*Hind*III fragment did not. A 3.6 kb *Hind*III fragment carrying homology was also subcloned from the cosmid into pUC18 as pDK1111. The 2.8 kb *Bam*HI-*Hind*III fragment was subcloned from pDK1110 into pUC18 and designated pDK1112. Complete restriction and Southern blot analysis of these three recombinant plasmids is currently being done since the results obtained with the *Sal*I and *Sma*I digestions preliminarily indicate that deletion constructions can be performed. Site-directed mutagenesis and enzymatic characterization of the mutants are planned in order to verify the isolation of *trpCD* from B. *japonicum*. Once verified, the corresponding DNA regions will be sequenced from strains I-110ARS, TA-11, and TA-11NOD$^+$ so that the molecular basis for alterations in tryptophan biosynthesis can be understood and manipulations can be performed.

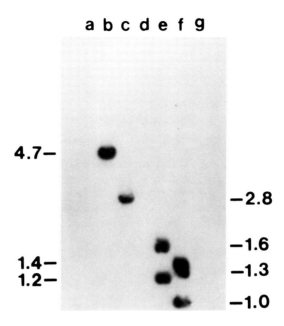

Figure 3. Southern blot analysis of a recombinant plasmid carrying B. japonicum DNA homologous to trpCD of R. meliloti. The agarose gel shown in Figure 2 was blotted and ^{32}P-labelled 4.7 kb EcoRI-PstI fragment carrying R. meliloti trpCD from pRmL104 was used as the probe. Molecular sizes indicated in the figure are in kilobase pairs (kb).

Discussion

Prototrophic revertants of a *B. japonicum* tryptophan auxotroph were found to promote better plant growth. This improvement in plant growth was due to increased nitrogen fixation per plant caused by an increase in nodule number. A new method of increasing nitrogen fixation in soybean plants by altering tryptophan biosynthesis in the bacterial microsymbiont and/or selecting for a prototrophic revertant of an ineffectively nodulating auxotrophic strain has been devised. Improvement of the nitrogen fixation ability of commercial bacterial inoculants for a wide range of crop legumes could conceivably result. Similar mutational alteration in the tryptophan biosynthetic apparatus of other rhizobia, transfer of the genes involved, and/or alteration of the host range of revertant strains could be useful in extending this method to other crops.

It was apparent that strains I-110ARS, TA-11 and TA-11NOD$^+$ are different in their tryptophan biosynthesis. This conclusion was supported by the following evidence. First, the TA-11 auxotroph, a well-defined tryptophan-requiring mutant, was known to be defective in a single enzyme of the tryptophan biosynthetic pathway (Wells and Kuykendall, 1983). Since TA-11NOD$^+$ revertants were obtained from strain

TA-11 in a single step selection, it is extremely unlikely that more than a single mutation was involved in the reversion to prototrophy. The change that occurred involved the tryptophan biosynthetic pathway as it corrected for the enzyme deficiency that existed in the auxotroph. Second, tryptophan in large amounts inhibits the growth of bradyrhizobia, and studies (Hunter and Kuykendall, 1990) showed that the wildtype progenitor strain I-110ARS was more sensitive to the effects of tryptophan than was TA-11NOD[+]. This higher resistance to tryptophan in the TA-11NOD[+] strain clearly distinguished it from the wildtype I-110ARS strain and was presumably due to an alteration in how the TA-11NOD[+] cells handle tryptophan biosynthesis.

The TA-11NOD[+] strain did not exhibit increased tryptophan catabolism. On the other hand, the tan 4b and 20d mutants of *B. japonicum* L-259 that exhibited enhanced nodulation and nitrogen fixation did so as a result of enhanced tryptophan catabolism (Kaneshiro and Kwolek, 1985; Kaneshiro, 1987; Kaneshiro and Nicholson, 1989). Increased resistance to tryptophan toxicity by the TA-11NOD[+] strain was clearly not due to enhanced tryptophan catabolism. Recent studies (Hunter and Kuykendall, 1990) have demonstrated that (1) strain TA-11NOD[+] did not exhibit enhanced tryptophan catabolism, and (2) strain TA-11NOD[+] was different and clearly distinguishable from the tan mutants of *B. japonicum* L-259.

Conclusions

Previously, tryptophan auxotrophs of *B. japonicum* were isolated and characterized. Of these, only mutants that were defective in tryptophan synthase, the last step of the tryptophan biosynthetic pathway, were capable of forming nodules. Recent research concerning the effect of tryptophan as a nutritional requirement of nodulation and symbiotic nitrogen fixation led to the discovery of a mutant strain having enhanced nodulation and symbiotic nitrogen fixation. Other mutants with defects earlier in the pathway formed abortive nodules as auxotrophs but prototrophic revertants did effectively nodulate. One mutant, deficient in indole glycerol phosphate synthase, produced prototrophic revertants that formed more nodules and fixed significantly more nitrogen per plant when compared to the progenitor wildtype strain I-110ARS. Heterologous probe DNA from *R. meliloti* was used to isolate *trpCD* from *B. japonicum*, and sequence analysis is planned.

References

Adams, T. H. , McClung, R. & Chelm, B. (1984) Journal of Bacteriology 159, 857-862.

Cole, M.A. & Elkan G.H. (1973) Antimicrobial Agents and Chemotherapy 4, 248-253.

Guerinot, M.L. & Chelm, B. (1986) Proceedings of the National Academy of Sciences USA 83, 1837-1841.

Hoch, S.O., Roth, C.W., Crawford, I.P. & Nester, E.W. (1971) Journal of Bacteriology 105, 38-45.

Hunter, W.J. (1987) Applied and Environmental Microbiology 53, 1051-1055.

Hunter, W.J. (1990) In: Nitrogen fixation: achievements and objectives, Gresshoff, P.M., Roth L.E., Stacey, G. & Newton, W.E. (eds.), Chapman & Hall, Publ. New York, London p 544.

Hunter, W.J. & Kuykendall, L.D. (1990) Applied and Environmental Microbiology 56, 2399-2403.

Kaneshiro, T. (1987) U.S. Patent #4,711,656.

Kaneshiro, T. & Kwolek, W.F. (1985) Plant Science 42, 141-146.

Kaneshiro, T. & Nicholson, J.J. (1989) Current Microbiology 18, 57-60.

Kaneshiro, T., Slodki, M.E. & Plattner, R.D. (1983) Current Microbiology 8, 301-306.

Kummer, R.M. & Kuykendall, L.D. (1989) Soil Biology and Biochemistry 21, 779-782.

Kuykendall, L.D. (1979) Applied and Environmental Microbiology 37, 862-866.

Kuykendall, L.D. (1987) In: Symbiotic Nitrogen Fixation Technology, Elkan, G.H. (ed.), Marcel Dekker, Inc., New York pp 205-220.

Kuykendall, L.D. & Weber, D.F. (1978) Applied and Environmental Microbiology 36, 915-919.

Leonard, L.T. (1943) Journal of Bacteriology 45, 523-527.

Norris, D.O. (1964) Commonwealth Bureau of Pastures and Field Crops Hurley Berkshire Bulletin 47, 186-198.

Shiio, I., Sato , H. & Nakagawa, M. (1972) Agricultural and Biological Chemistry 36, 2315-2322.

Vincent, J.M. (1970) A Manual for the Practical Study of RootNodule Bacteria. Blackwell Scientific Publications, Oxford.

Wells, S.E. & Kuykendall, L.D. (1983) Journal of Bacteriology 156, 1356-1358.

Alfalfa nodule development; ribonucleotide pools and ribonucleotide reductase activity in cultured and symbiotic *Rhizobium meliloti*

J. R. Cowles and S. F. Yet

Dept. of Biology, Vir. Polytechnic Institute & State Univ., Blacksburg VA 24061-0308

When alfalfa plants (*Medicago sativa*) are infected with *Rhizobium meliloti* the resulting nodules grow longitudinally after they emerge from the alfalfa roots. In young (2-4 week-old) nodules two distinct regions can be distinguished on the elongating nodules. The white tip composes 10 to 15% of the nodule length while the remainder of the nodule is larger in diameter and flesh colored. The tip region is where host cell division occurs and where these cells are invaded by infection threads and become infected with rhizobia. Many host cells in the flesh colored region are essentially filled with rhizobia in various stages of development. Large amounts of leghemoglobin are synthesized in this region which accounts for its characteristic color. As the nodule continue to grow and age, a third region emerges at the point of root attachment. This region is typically brown in color. The brown (basal) region becomes increasingly longer with nodule age, often exceeding 50% of nodule length in 14- to 16-week-old nodules. The basal region apparently is the senescence region of the nodule for both host cells and bacteroids. Leghemoglobin also is being degraded in the basal region. More than 95% of dinitrogen fixation in nodules occurs in the flesh colored region subsequently referred to as the middle region.

The three nodule regions described above, tip, middle and basal, allow for convenient spacial separation of alfalfa nodules into different stages of development and function (Paau, Cowles and Raveed, 1978). The bacteroids from these regions have been described with respect to relative size, and nucleic acid content (Paau and Cowles, 1978). Bacteroids in the nodule tip are similar in size and DNA content to free-living (cultured) *R. meliloti*. Cells in the middle region contain bacteroids with a range of sizes including a significant population of large (5 to 7 μm) bacteroids. The average DNA content in this population of bacteroids is considerable greater than those located in the tip region. Bacteroids in the basal region also exhibit a range of sizes but are slightly larger than bacteroids in the middle region. Interestingly, DNA content in bacteroids of the basal region is much smaller than in those of the middle region. This is evidence that bacteroids in this region are in senescence.

DNA synthesis is obviously essential for continuous cell growth and division. In alfalfa nodules the rhizobia grow and divide to the extent that they essentially fill many of the host cells. Some bacteroids, however, become abnormally large accumulating increased levels of protein, nucleic acids etc., but apparently are unable to divide under the limiting environment of the nodule. Why are these large bacteroids formed and what are their function in alfalfa nodules? As noted above the large bacteroids have increased amounts of DNA which suggest active DNA synthesis in these cells. Rhizobia have been shown to contain very high levels of ribonucleotide reductase, Cowles, Evans and Russell (1969) a regulatory enzyme in DNA synthesis in comparison to most other organisms including rapidly growing *E. coli* (Reichard, 1962). Ribonucleotide reductase in *R. meliloti* requires B_{12} coenzyme for activity. Whether this rather unique property is important in establishing symbiosis is not known. An interesting characteristic of ribonucleotide reductase is that reduction of all four common ribonucleotides is catalyzed by the same enzyme and that catalytic activity is highly regulated (Cory and Sato, 1983). The most common allosteric regulators are deoxyribonucleotides. Certain nucleotides are positive allosteric effectors and others are negative effectors. Ribonucleotide reductase activity in alfalfa bacteroids is about 20% of that in cultured *R. meliloti* but still high in comparison to most organisms. The highest levels of ribonucleotide reductase activity in bacteroids are found in bacteroids isolated from the middle nodule region. Ribonucleotide reductase activity in bacteroids of the middle regions was 6- to 7- and 4- to 5-fold higher than in bacteriods of the tip and basal regions, respectively. The difference in ribonucleotide reductase activity between bacteroids from the middle and tip regions suggest a positive correlation between ribonucleotide reductase activity and bacteroid size. The substantial decline in ribonucleotide reductase activity in bacteroids of the basal region relative to these of the middle region again suggest these bacteroids are in senescence. To further examine correlation of ribonucleotide reductase activity to bacteroid size, bacteroids in the tip and middle regions were fractionated into populations enriched in large and small bacteroids. Ribonucleotide reductase activity in the enriched large bacteroid samples were 85-fold greater than that of the small bacteroid samples. These results again demonstrate a strong positive correlation between ribonucleotide reductase activity and bacteroid size.

Since ribonucleotide reductase activity is regulated by nucleotides then relative size of the nucleotide pools should influence ribonucleotide activity in *R. meliloti*. To examine this possibility we (1) determined the nucleotide pools in cultured and symbiotic *R. meliloti* (2) looked for nucleotides that had unusual quantitative changes and (3) examined what affect nucleotides of interest had on ribonucleotide reductase activity. In the particular study we focused on ribonucleotide rather than deoxyribonucleotide pools.

For this study *R. meliloti* cultures were grown on mannitol media and after harvesting the cells were suspended in a phosphate buffer solution and extracted with perchloric acid. After centrifugation the supernatant containing the ribonucleotides were separated and quantitated by high performance liquid chromatography. The optimum chromography conditions were established using 12 ribonucleotide standards; mono-, di-, and tri-nucleotides of adenine, quinine, cytosine and uridine bases. Analysis of cultured rhizobia grown to stationary phase contained 10 of the common ribonucleotides plus three peaks identified as monosaccharide bound dinucleotides (Fig. 1). The most abundant ribonucleotide in

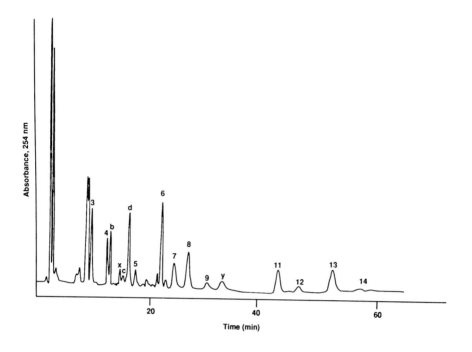

Figure 1. Ribonucleotide profile of cultured R. meliloti. *The cultures were harvested 28 hours after inoculation (stationary phase). Ribonucleotide separation was by HPLC using an ion exchange column (Partisil-10 SAX, 0.46 x 25 cm, Whatman). The ribonucleotides were eluted with a linear gradient; Buffer A contained 7 mM KH_2PO_4, pH 4.0 and Buffer B 250 mM KH_2PO_4, pH 4.5 plus 0.5 m KCL. The labelled peaks were identified as follows: Peak 3, UMP; 4 GMP, 5 XMP 6, UDP; 7, CDP; 8, ADP; 9, GDP; 11, UTP; 12 CTP; 13, ATP; 14, GTP, b UDP-galactose; C, ADP-ribose; d, GDP-mannose, X and Y, unknowns.*

the rhizobium extracts was UDP. In fact, the uridine nucleotides as a group composed about 50% of the total ribonucleotide pool in R. meliloti. Cytosine nucleotides were the least abundant group. Interestingly, the amount of GDP-mannose was about 4-fold higher than free GDP. The other two bound dinucleotides, ADP-ribose and UDP-galactose were present in smaller amounts that the free nucleotides.

The ribonucleotide pools also were quantitated in R. meliloti cultures grown to different growth stages (ages). Overall size of ribonucleotides pools were greatest in late log cultures and smaller in both younger and older cultures. Most but not all individual ribonucleotides followed this pattern. The principal exceptions were cytosine nucleotides in which pools were lowest in late log cultures. Of the cytosine group, CDP exhibited the most dramatic concentration change. The change in CDP pools relative to other dinucleotides at different culture age is shown in Fig. 2. The

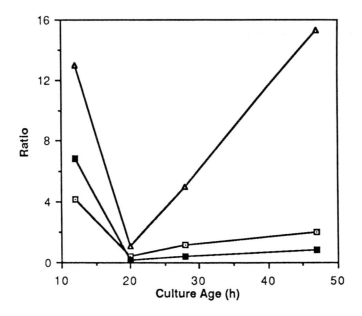

Figure 2. Comparison of ribonucleoside diphosphates ratios in cultured R. meliloti *grown to different culture ages. CDP/ADP (–□–) CDP/UDP (–■–) CDP/GDP (–Δ–). The ribonucleoside diphosphate ratios were calculated from ribonucleotide profiles of different culture ages, one of which is shown in Fig. 1. Culture ages were selected to represent different stages of rhizobia growth; 12h-early log phase, 20h-late log phase, 28h-stationary phase, 47-late stationary phase.*

most dramatic difference is the rapid decrease in CDP ratios between early and late log cultures. The only other nucleotides that exhibited substantial change was a 4-fold increase in UDP/ADP ratios between early log cultures and all older ones.

The major ribonucleotides also were detected in cell-free extracts of alfalfa nodule bacteroids. In bacteroids, ATP was the most abundant ribonucleotide. UDP and UTP pools were much smaller in bacteroids than in cultured rhizobia. The size of ribonucleotides pools in different aged nodules followed a pattern similar to that of cultured rhizobia; largest pool size in rapidly growing nodules and smaller pools in both younger and older nodules. The CDP/NDP ratios also were lowest in rapidly growing nodules which is consistent with that observed in cultured R. *meliloti*. The

CDP/NDP ratio differences were even more dramatic when compared in small and large bacteroids. The CDP/NDP ratios, in small bacteroid fractions were 50 to 60-fold larger than in large bacteroid fractions. These results along with data in Fig. 2

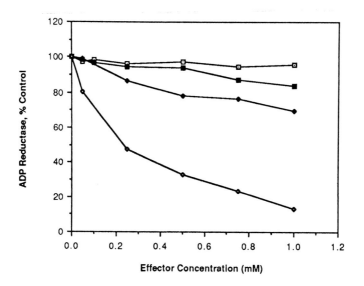

Figure 3. Effect of CDP and uridine ribonucleotides on ribonucleotide reductase activity (ADP reduction). CDP (–●–), UMP (–□–), UDP (–◆–), UTP (–■–). The control reaction mixture contained in a final volume of 0.5 ml; 50 mM potassium phosphate, pH 7.3; 30 mM dihydrolipoiate; 20 mM coenzyme B_{12}; 1 mM EDTA-Na$_2$, 0.25 mM ADP (substrate) and 50 μg partially purified enzyme. In reactions containing effectors, CDP, UMP, UDP or UTP were added at indicated concentrations. Reactions were incubated at 37 °C for 30 min then placed in an ice bath. After treatment with chloroacetamide the deoxyribonucleotide content was determined by dipherylamine reagent (control activity was 770 nmoles in protein^{-1} . h^{-1}).

demonstrate an inverse correlation between CDP/NDP ratios, and cell growth and ribonucleotide reductase activity.

CDP may have a unique regulatory role based on its change in pool size relative to most other ribonucleotides and its inverse correlation to rhizobia growth. One possibility is that CDP may be a regulator of ribonucleotide reductase activity. Typically, ribonucleotide reductase activity is regulated by deoxyribonucleotides but a couple of ribonucleotides also have been demonstrated to effect enzyme activity. CDP and uridine mono-, di-, and tri-nucleosides were tested as possible effectors of in vitro ribonucleotide reductase activity. CDP at concentrations of 0.25 mM decreased ADP reduction 50% while the uridine mono- and tri-nucleosides had much smaller effects on ADP reduction (Fig. 3). Ribonucleotide reductase activity decreased about 20% at 1.0 mM UDP concentrations whereas ADP reduction decreased about 85%. These same ribonucleotides were tested as potential effectors of GDP reduction. CDP and UDP affected GDP reduction with 35 to 45% reduction of activity at 1.0 mM concentrations.

The results from Fig. 3 were re-plotted to establish the type of inhibitory effect exhibited by CDP and UDP. A Dixon plot analysis of the data shows that CDP inhibited ADP reduction non-competitively while UDP was a weak competitive inhibitor of ADP reduction. This suggest that UDP is competitive with ADP at the catalytic site while CDP inhibits ADP reduction primarily through its action at an effector site. The same analysis of UDP and CDP inhibition of GDP reduction showed both dinucleotides to be competitive with GDP.

Large bacteroids in alfalfa nodules are an interesting mystery. They are not a requirement for dinitrogen fixation per se since they are not present in effective dinitrogen fixing systems such as soybeans. Then why do they exist and what is their function in alfalfa? One possible reason is that they are controlled forms of *R. meliloti*. For free-living microorganisms to establish a symbiotic association with a host requires proper controls to keep the microorganisms functional but yet unable to grow and divide to the extent that they would becomes parasitic or pathogenetic. Perhaps large bacteroids in alfalfa nodule is the controlled form of *R. meliloti*. In these cells, nucleic acid, proteins etc. accumulate in preparation of cells prior to cell division but for some reason the cells are unable to divide. Another interesting question with regards to the large bacteroid is to what extent are they capable of dinitrogen fixation? We assume they are capable of dinitrogen fixation but are they more or less effective than the smaller bacteroids present in the same host cells?

Results from the present study demonstrates that large bacteroids have much higher (85-fold) *in vitro* ribonucleotide reductase activity than small bacteroids. This suggests that large bacteroids have increased amount of at least this critical enzyme in DNA synthesis to support the increased levels of DNA. In addition, the quantity of CDP which non-competitively inhibits ribonucleotide reductase activity is drastically lower (45-fold) in large bacteroids as compared to small bacteroids. This provides a more favorable environment for ribonucleotide reductase to function. The changes in ribonucleotide reductase activity and CDP quantities are consistent with a working hypothesis that the large bacteroids in the middle nodule region are synthesizing large quantities of DNA.

References

Cory, J.G. & Sato, A. (1983) Molecular and Cellular Biochemistry 53/54, 257-266.

Cowles, J.R., Evans, H.J. & Russell, S. H. (1969) Journal Bacteriology 97, 1460-65.

Paau, A.S., & Cowles, J.R. (1978) Canadian Journal Microbiology 24, 1283-1287.

Paau, A.S., Cowles, J.R. & Raveed, D. (1978) Plant Physiology 62, 526-530.

Reichard, P. (1962) Journal Biological Chemistry 237, 3513-3519.

Systemic regulation of nodulation in legumes

Peter M. Gresshoff and Gustavo Caetano-Anollés

Plant Molecular Genetics and Center for Legume Research, The University of Tennessee, Knoxville, TN 37901-1071. USA

Introduction

Plants, like other multicellular organisms, possess the ability to regulate the growth and development of individual cells or tissue structures. This regulation occurs by a variety of mechanisms as both short and long distance interactions are required. Short distance regulation is akin to "cell sociology", in which physical contact, diffusion of small molecules (such as oxygen, ammonia, or even protons), and surface receptors interact to modulate gene expression and cell multiplication to form cellular arrangements and organs. Long distance communication may involve source/sink relationships of common metabolites (such as glutamine, sucrose, glucose), or specific compounds (such as oligosaccharins or phytohormones), which transmit morphological signals and result in differential gene expression and development. The gradient or concentration of the signal substance itself may not be sufficient or may not contain the necessary information for the developmental response; a differential sensitivity may result from the chemical nature of the signal substances, differing receptor concentrations, or differing receptor types. When discussing systemic communication in plants one is immediately attracted to the possible role of phytohormones, which are substances that are produced by the plant and which at low concentrations result in a developmental response at sites usually distant to their site of synthesis. Plant hormone biology differs from that of the animal kingdom, as animals use well defined glands, and hormones are often clearly definable peptides, amino acid analogues, lipids, or steroids eliciting more or less specific responses in target tissues. Plant researchers for decades have attempted to correlate their findings with those in animals by proposing that specific plant tissues are the source tissues of specific phytohormones.

This article is dedicated to Ben B. Bohlool who lost his life on May 16, 1991 in Maui. His contributions to the subject were pivotal and seminal.

The research involving auxin synthesis and the classical *Avena* coleoptile curvature experiments have been of relevance in this respect. Other examples of specific sites of synthesis are numerous; e.g. the quiescent center behind the growing root tip as a source of cytokinin biosynthesis, the aleurone layer of barley as the source of gibberellin release that then stimulates amylase production in the endosperm, and the apical shoot tip producing auxin (for a comprehensive analysis of the subject, see Letham et al, 1978).

However, progress in plant biology is hindered by our meager understanding of fundamental aspects of development. Few pathways involving phytohormone biosynthesis are understood. Little is known about receptor proteins and signal response chains. Underlying developmental mechanisms as those involved in cell wall thickening or their localized maturation are poorly described. Even the genetics of plant development has not achieved the level of study seen in *Drosophila*, nematode, or slime mold development, despite significant advances in the last decade especially with model plants such as *Arabidopsis thaliana*. The analysis of systemic responses governing nodulation that involve cell division, expansion and differentiation is similarly curtailed. Viewed in this way nodulation research takes on an extra perspective beyond its agricultural focus, as the analysis of nodule initiation allows investigation of plant development.

The legume nodule is a complex organ as shown by its histological and physiological study in different plant species. Nodules in general contain infected and uninfected cells (Gresshoff and Delves, 1986). There are vascular bundles with their component tissues as well as nodule parenchyma and endodermis. Young nodules and those of the indeterminate type possess meristematic regions in which cell division is maintained. Various biochemical processes can be detected by the presence of new proteins and enzymatic activities absent in the original root tissue. Nodules function as a carbon and nitrogen sink (Hansen et al, 1990), and if invaded with compatible *Rhizobium* or *Bradyrhizobium*, they act as a nitrogen source. Specialized enzymes are induced or activated in developing nodular tissue (see Mellor and Werner, 1990). These often fall into the category of 'nodulins', which are defined as nodule-specific or amplified proteins (Franssen et al, 1990; Nap and Bisseling, 1990).

The whole plant regulates both the development and functioning of nodules. The subject of this article focuses on how regulation of nodule development operates. Although there are numerous versions of legume nodules (see Sprent et al, 1990), studies were mainly restricted to nodules from major crop legumes, such as soybean, clover, pea, and alfalfa. In general we recognize two major nodule types: those with a persistent meristem, called indeterminate or meristematic, and those without such a meristem, called determinate. Early cross-inoculation studies showed that nodule structure is controlled by the plant genome and not the bacterium. One of the most dramatic examples of this fact comes from the extraordinary case of *Parasponia* nodulation (Gresshoff et al, 1985; Scott and Bender 1990). *Parasponia* is a non-legume which is nodulated effectively by *Bradyrhizobium* strains like ANU289 and CP283, which also nodulate the forage legume siratro (*Macroptilium atropurpureum*). While nodules on siratro are determinate like soybean nodules, nodules on *Parasponia* resemble modified lateral roots, being coralloid with a central vascular system (Price et al, 1984). Since the same bacterium induces nodules of different basic morphology in different hosts the host genome must specify nodule morphology. This concept is strongly supported by the discovery of spontaneous nodules formed in the absence of *Rhizobium* on specific genotypes of alfalfa (Caetano-Anollés et al, 1991; also this volume).

The number of nodules per plant is limited and generally not restricted by the availability of infective bacteria. For example, nodulation of legumes in the field and under laboratory inoculation conditions is generally confined to the root crown, leaving lower root portions relatively non-nodulated (Allen and Allen, 1981). Yet these lower regions are able to nodulate if the plant is inoculated in a delayed fashion. There is no plant ontogenetic barrier to nodulation and it appears as if earlier nodules suppress the development of younger nodules (Pierce and Bauer, 1983; Bauer, 1981). From these observations the concept of autoregulation or feedback regulation of nodulation was born.

Nitrate inhibition of nodulation

The legume root nodule symbiosis is also controlled by nitrate (Carroll and Mathews, 1990). Nitrate inhibition of nodulation was assumed to be caused by a metabolic and nutritional balance in which the nitrogen status of the plant would shift from that poised for symbiotic nitrogen acquisition to that of nitrate reduction. This model was probably based on similar metabolic regulation in bacteria, where for example glucose or ammonium regulate the utilization of alternative sugars or nitrogen sources, respectively. However, the regulation of nodulation by nitrate is a much more complex process that involves other steps in early nodule development as will be discussed later.

Interestingly, the inhibitory effect of nitrate is also localized (Hinson, 1975; Carroll and Gresshoff, 1983). Concurrent exposure of one portion of a split root system to nitrate and *Bradyrhizobium* and the other half to *Bradyrhizobium* alone gave significant inhibition of nodulation and nodule development in the nitrate exposed tissue, but no inhibition in the nitrate-free root system. However, specific nitrogenase activity was inhibited equally on both sides, showing that the effect on nitrogenase was systemic. Since short-term nitrate effects on nitrogenase activity are mediated through regulation of the variable oxygen barrier existing in the inner cortex of the nodule (Layzell et al, 1990), we postulate that the signals for the initiation of the oxygen barrier also travel systemically.

Autoregulation of nodulation

It was only the work of Phillip Nutman that uncovered quantitative relationships between nodule size, distribution, and number. If a plant had few nodules, they were usually large. If there were many, they were small. Seemingly plants possessed internal homeostatic controls that judged the amount of nodule tissue. The nodule meristem itself was important for the regulation of nodulation (Nutman, 1952). Removal of nodule meristems from clover plants that had achieved a constant nodule number resulted in a renewed onset of nodulation. Nutman tried to clarify the contribution of the plant to nodule development by isolating symbiotic mutations in red clover. Although these were novel in concept, their utility was limited because of the absence of concomitant genetic and biochemical studies.

Recently the nodule removal experiment were repeated with both soybean (Caetano-Anollés et al, 1991b) and alfalfa (Caetano-Anollés and Gresshoff, 1991b,c) to show quantitatively, and in both time and space, that these two species respond differently. Soybean developed new nodules in the region of initial nodulation in the primary root (i.e. within the original window of nodulation, i.e. the region between the root tip and the emerging root hairs at the time of inoculation). Such nodules harbored the original bacterial inoculum and not a superinfecting strain added at the time of nodule excision. This suggests that soybeans, and perhaps all other tropical legumes with determinate nodules, arrest nodule primordia during early nodule ontogeny, and that this arrest is maintained by nodules that were developed first. Such mechanisms are similar to apical dominance, where the first-formed organ suppresses similar organs in the same region.

In contrast, nodules formed in alfalfa in the root tip regions expected to be the site of new nodule formation (Caetano-Anollésand Gresshoff, 1991c). None were found in the initial window of nodulation. Alfalfa does not arrest a major proportion of its nodule primordia at an intermediate stage, but regulates the establishment of pre-nodules by a yet unknown mechanism during the onset of cell division. This is reflected by the speed of systemic suppression in split roots (Caetano-Anollés and Bauer, 1988) and the location of nodules in the root system.

These conclusions were confirmed by histological analyses, which showed that (a) soybean roots contained abundant early nodulation stages (Calvert et al, 1984; Mathews et al, 1989a; Gerahty et al, 1990, 1991), and (b) alfalfa roots contained only a small number of undeveloped nodule primordia (Caetano-Anollés and Gresshoff, 1991c), showing that autoregulation occurs earlier in alfalfa than soybean. Whether the distinction between soybean and alfalfa is the result of the tropical-temperate legume difference, represents coincidence, or the fundamental response of the nodules of meristematic and determinate type was not tested. However, a common feature is that autoregulation in both soybean and alfalfa arrests pre-nodule stages through a control of cell division, and that the control point lies prior to the onset of nitrogen fixation and does not depend on the nitrogen status of the plant (see Caetano-Anollés and Gresshoff, 1990 for a model).

Nutman's research was advanced by the work of Bauer and co-workers three decades later (Bauer, 1981; Bhuvaneswari et al, 1980; Pierce and Bauer 1983; Calvert et al, 1984). The analysis of nodulation was simplified by using plants grown in plastic pouches and investigating the appearance of nodules along the primary root, relative to the position of the growing root tip and the zone of emerging root hairs at the time of inoculation. Only the region close to the root tip (RT) at the time of inoculation was susceptible to infection. Mature root hairs (in general) were resistant to infection. Prior inoculation of a region of the root prevented further nodulation in younger region of the main root, even if accompanied by a second inoculation. Accordingly newly developing root tip regions, although susceptible to infection, were not equally able to develop nodules. Since the change in the ability to form nodules in the newly developing root tips was dependent on prior inoculation of the plant and occurred relatively rapidly, non-emergent nodules had to influence the fate of pre-nodules in the ontogenetically younger tissues. This analysis constituted the first quantitative description of autoregulation. Since then, evidence for similar autoregulation was described in siratro (Ridge and Rolfe, 1985), subterranean clover (Sargent et al, 1987), *Parasponia* (Bender and Rolfe, 1988) and alfalfa (Caetano-Anollés and Bauer, 1988).

Inhibition of nodulation may be caused by linear transmission of an inhibitor or the exhaustion of a critical substance by the earlier developed nodules. This question was resolved by the elegant experiments of Kosslak and Bohlool (1984), using soybean split root systems separately inoculated in time and space. Prior inoculation of one root portion systemically inhibited nodulation in the other. Thus the signal controlling nodulation was very likely to involve the shoot. They also found that plants grown under different light regimes gave different levels of suppression.

The analysis of the physiology of this systemic communication is complex. Many molecules like sucrose, phytohormones, and nitrogenous compounds are transported both symplastically as well as in the vascular stream. The extent of distribution and targeting of such translocated molecules seems to be influenced by specific physiological conditions. Nitrate as well as infection of a certain root portion of a split root system affected the translocation of photosynthate (structural resources) to a treated root portion (Singleton and van Kessel, 1988).

Olsson (1988) investigated the interaction of exogenously supplied phytohormones, nitrate supply, and nodule number per plant with wild-type and supernodulating soybeans (Table 1). In experiment A, where indole acetic acid (IAA) was injected daily into the stems of soybean plants, she noted that 1μM IAA suppressed nodulation in wild-type Bragg by 60% compared to buffer injected controls. If high nitrate was supplied the nodulation in the control was lowered by about 60% and concomitant injections of otherwise inhibitory concentrations of IAA had no effect. A similar behavior was observed with gibberellic acid (GA_3). The conclusion was that the nitrate status of the Bragg plant influenced whether injected IAA or GA_3 inhibited nodulation. When the experiment was repeated with supernodulation and nitrate tolerant mutant nts382, only slight inhibition was observed in both nitrate treatments. In contrast, the injection of GA_3, but not IAA, inhibited nodule numbers in nts382 independently of the nitrogen status. In a separate experiment abscisic acid (ABA) was injected into plants growing at high nitrate concentrations (Table 1, experiment B) . High concentrations of ABA (50μg per injection) inhibited nodulation in both genotypes, although intermediate levels (i.e. 1μg and 5μg) resulted in differential inhibition of nodulation. The mutant nts382 injected with 1μg and 5μg ABA had 88% and 49% of the nodulation levels of the control respectively, while corresponding values for Bragg were 12% and 8%. Table 1 (experiment C) shows another experimental approach. GA_3 and its biosynthesis inhibitor chloro-choline-chloride (CCC) were sprayed onto plants under different nitrate treatments. The control data illustrated the strong inhibition of nodulation in Bragg and the nitrate tolerance in nts382. While the GA_3 treatment lowered nodulation in all cases (not as significantly as in the micro-injected tests shown in experiment A (Table 1), CCC gave significant increases in nodule number per plant.

Table I: Phytohormone effects on nodulation of soybean Bragg and mutant nts382

Treatment	Bragg	Bragg (+KNO₃)	nodule number per plant nts382	nts382 (+KNO₃)
Experiment A:				
control	63 (9)	23 (3)	408 (41)	521 (92)
IAA 1μM	24 (5)	31 (17)	361 (37)	384 (73)
IAA 10μM	25 (8)	27 (15)	276 (61)	176 (53)
GA₃ 1μM	24 (7)	35 (7)	141 (46)	393 (126)
GA₃ 10μM	15 (5)	8 (2)	146 (47)	175 (71)
Experiment B:				
control	nt	49 (8)	nt	949 (186)
ABA 50 ng	nt	44 (8)	nt	1100 (294)
ABA 500 ng	nt	38 (13)	nt	1011 (213)
ABA 1 μg	nt	6 (3)	nt	834 (12)
ABA 5 μg	nt	4 (4)	nt	465 (83)
ABA 50 μg	nt	0 (0)	nt	3 (4)
Experiment C:				
control	187 (33)	22 (6)	587 (55)	459 (88)
GA₃ 3 μM (spray)	159 (40)	15 (6)	391 (83)	183 (85)
CCC 63 μM (spray)	256 (39)	82 (19)	1016 (287)	776 (128)

Groups of 7 to 12 plants were either grown under low nitrate conditions (0.5 mM KNO_3, a level known not to interfere with nodulation) or 5.5 mM KNO_3. Plants were injected into the intercotyledonary node with 5 to 10 μl of a solution containing the different chemicals (or water controls). Variable amounts of liquid were retained in injected tissues leading to experimental variability. Plants were inoculated with *Bradyrhizobium japonicum* strain USDA110 generally 9 days after planting and injected from that day on for a further 12 days. Plants in experiment C were treated by daily foliar spray to run off. Data are means with standard deviation in parentheses and are taken from Olsson (1988), nt= not tested.

These findings support the claim that phytohormones are involved in the regulation of nodulation, but that their function is complex and at times modulated by other physiological parameters such as the nitrogen status of the plant.

It is tempting to compare the nodulation systemic response to wound-induced systemic responses. Studies on wounding have shown that physical or microbial injury to a leaf will induce wound response proteins in other leaves of the plant. Ethylene and abscisic acid are proposed to be intermediate messengers. However, other molecules may be utilized. For example, Farmer and Ryan (1990) discovered methyl jasmonate as an alternative messenger for the systemic wound response.

Autoregulation mutants

About a decade ago it became obvious that the analysis of the nodule symbiosis required the extensive analysis of its genetic components (see Vincent, 1980, for a trend-setting chapter and perspective). This was quickly achieved with bacteria, as transposon mutagenesis and gene isolation defined a set of nodulation and nitrogen fixation genes. Today the function of some of these genes is being understood (Lerouge et al, 1990; Long et al, 1991; Truchet et al, 1991). A parallel development occurred with the genetic analysis of plants, but using the two different perspectives of molecular biology and genetics.

Verma and co-workers were the first to focus on the tissue-specific expression of plant genes during nodule development (Legocki and Verma, 1980). Their work developed the concept of nodulins, proteins that are specifically expressed in the legume nodule. Today we know that this exclusive definition may have been too restrictive as molecular evidence has accumulated showing that some nodulin probes detect expression in other plant organs such as the flower (Franssen et al, 1990). We must, however, be careful about such interpretations as the possibility for multigene families exists, with specific members being expressed in specific organs. The members of the multigene family could share sufficient homology to make distinction difficult.

Their studies produced fundamental results on leghemoglobin (Lee and Verma, 1984), uricase (Nguyen et al, 1985), sucrose synthetase (Thummler and Verma, 1987) as well as several nodulins associated with the peribacteroid membrane (Fortin et al, 1985, 1987; see also Roberts et al, this volume). In summary, the early research focussed on gene expression in established nodules and detected nodulins involved in nitrogen fixation and ammonia assimilation rather than in nodule development.

Extensive work from the laboratory of Bisseling (reviewed in Sanchez et al, 1991; Nap and Bisseling, 1991; Caetano-Anollés and Gresshoff, 1991d) concentrated on early nodulins expressed prior to the onset of nitrogen fixation. Several of these nodulins were discovered, but ENOD2, a hydroxyproline-rich cell-wall protein related to extensin, was the most studied (Franssen et al, 1987). Homologous copies of the enod2 gene were found in other legume systems (Kouchi et al, 1989, Trese and Pueppke, 1990; Dickstein et al, 1988; Norris et al, 1988), suggesting its universal involvement in nodule development. Nodulin research has defined several genes

and their control regions. With the advent of legume transformation it was possible to transform chimeric nodulin promoter-reporter gene constructs into recipient plants and monitor the expression of these genes under different environmental and physiological conditions (Diaz et al, 1989; Verma, 1990; Debruijn et al, 1990; Bogusz et al, 1990). Recently, *enod2* induction was shown to occur rapidly after exposure of roots to cytokinin (Debruijn et al, 1990). Further attempts to clarify functions of unknown nodulin regulatory regions rely on sequence comparisons to known genes or the potential expression of cloned genes and biochemical characterization in *Escherichia coli* as shown by Suzuki and Verma (1991) for nodulin 26.

An alternative approach to "classical" molecular biology was the genetic study used by our laboratory for the soybean system. Several plant gene loci controlling nodulation were found by chemical mutagenesis. While the genetic, biochemical and physiological characterization of these plants is now relatively advanced, little is known about the genes themselves. Because gene products may be present at very low levels and thus molecular techniques such as cDNA libraries or *in vitro* translation product analysis are prone to fail, we chose to isolate these genes by reverse genetics (see Funke and Gresshoff, 1991; Landau-Ellis et al, 1991, and this volume).

A related genetic study using restriction fragment length polymorphisms (RFLPs) was used by LaRue and co-workers to map nodulation deficiency genes in pea (Weeden et al, 1990). With this approach, they found that several nodulation related gene loci were clustered.

One significant mutant type was affected in the autoregulation response (Carroll et al, 1985a,b) resulting in profuse nodulation. These supernodulation mutants in soybean also possessed a nitrate tolerant nodulation (nts) phenotype, suggesting a shared mechanism involving internal and external nodule regulation (Day et al, 1989). Supernodulation mutants were also discovered by others in soybean using our procedures (Gremaud and Harper, 1988; Buzzell et al, 1990; Akao and Kouchi, in preparation). Several other supernodulating mutants were also found in pea (Duc and Messenger, 1989; Jacobsen, 1984).

The predominant research on soybean supernodulation was carried out with line nts382. This plant line is able to develop as many as 4000 nodules per plant (D. Day and A. C. Delves, pers. comm.). To study the possible systemic communication in an autoregulatory mutant plant, reciprocal grafts in either the hypocotyl or epicotyl region were used to demonstrate that the supernodulation phenotype was controlled by the genotype of the shoot portion (Fig 1A; Delves et al, 1986). In other words, a grafted plant with a wild-type root system but with the shoot from nts382 nodulated like a supernodulating control. This observation was true for all tested nts mutants of soybean (Delves et al, 1987a; Buzzell et al,1990; Akao and Kouchi, 1991). Grafts involving the subspecies *Glycine soja* and *G. max* nts382 showed that even *G. soja* used the shoot for autoregulation (Delves et al, 1987b). Experiments in which shoot and lateral apexes were removed after reciprocal grafting between wild-type shoots and nts roots suggest that the leaf, and not the growing tip of the shoot, has a major role in the autoregulation circuit (Delves et al, 1991).

The question remained: how is the shoot (or leaf) of an inoculated plant able to sense the inoculation event? While the split root experiments of Kosslak and Bohlool (1984) revealed the systemic nature of initial nodule suppression, and Olsson et al (1989) showed that the nts382 mutant lacked the severe systemic suppression (Fig 1B), both studies did not address the source of the triggering. Caetano-Anollés and Gresshoff (1990) extended the split root system by using approach grafts between plants of differing genotypes (Fig 1C). In this culture system, plants stems were cut longitudinally and fused to heal, resulting in plants with two shoots and two roots. The root portions were inoculated separately in time and

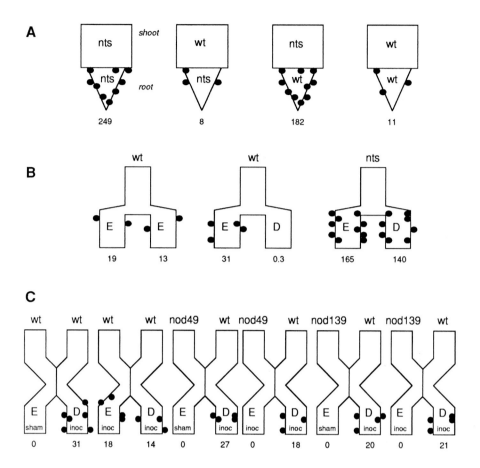

Figure 1: Experimental designs to test systemic signalling in soybean: Panel A (Delves et al, 1986): nts = mutant nts382, wt= wild type cv. Bragg. Plants were grown uninoculated prior to grafting. Inoculation with B. japonicum *strain USDA110 occurred 4 days after reciprocal grafting in the hypocotyl region. Panel B (Olsson et al, 1989): Split root systems were inoculated at an early (E) time point or one week (D) later. Panel C (Caetano-Anollés and Gresshoff, 1990): nod49 and nod139 are nonnodulating soybeans. The left hand portion of the approach-grafted plants was used as a sensor of inoculation responses, while the right portion was used as a reporter. The inoculation was with strain USDA110, either early (E, 4 days after grafting) or 72 hours later (D). "Sham" means that no bacteria were added. "inoc" signifies inoculation. Filled circles represent nodules and nodule number per root portion is indicated.*

space, as in split root experiments. In contrast to split roots the approach grafts allowed the generation of chimeric plants containing two root genotypes. When Bragg (wild-type) was grafted to itself and inoculation on the second root portion with *Bradyrhizobium japonicum* strain USDA110 was delayed by 72 hours after inoculation of the first root portion, a systemic suppression response was observed. This result is consistent with the systemic suppression found in split root plants (Olsson et al, 1989). About 50% fewer nodules developed on the second root portion. This experimental set-up allowed the use of the first inoculated side as the 'sensor' of the inoculation response, while the delayed inoculated side was used as the 'reporter'. Mathews et al (1989 a,b) characterized two non-nodulation mutants of soybean cv. Bragg that were previously obtained by Carroll et al (1986). The mutants nod49 and nod139 were caused by recessive mutations in two separate genetic loci. While they were not induced to curl root hairs, nod49 was capable of some cortical cell division but nod139 was not. When nod139 was used as the sensor portion in approach grafts, no suppression of nodulation occurred on the wild-type reporter side (Fig. 1C). When nod49 was the sensor root, suppression similar to that for wild-type roots was observed on the reporter side. These findings were used to extend the current model of systemic communication in autoregulation (Gresshoff and Delves, 1986) by including the possible source of the root signal (Caetano-Anollés and Gresshoff, 1990). Cortical cell divisions or developmental events closely coupled with them were proposed to produce a signal which is translocated to the leaf to initiate autoregulation. The leaf then generates response signals to inhibit further sustained cell division in the root cortex.

The analysis of cortical cell division stemmed from the original anatomical work of the late Calvert and his co-workers, who discovered pseudo-infections, defined as division clusters without bacterial penetration that are unable to form mature nodule meristems (Calvert et al, 1984). Mathews et al (1989a) used similar serial sectioning of Bragg and nts382 to show that the initially susceptible region of the primary root possessed similar numbers of infection stages, but that Bragg had an abundance of early stages while nts382 plants permitted an increased transition of these beyond stages IV and V (as defined by Calvert et al, 1984). nts382 appeared altered in its ability of blocking the development of nodules beyond stage IV. More recently, Gerahty et al (1991) confirmed and extended these results, illustrating complex kinetic and spatial controls. Bragg roots developed stage IV pre-nodules within 48-72 hours. The analysis of split root systems (Olsson et al, 1989) and nodule distribution profiles (Caetano-Anollés and Gresshoff, 1991a) point to an onset of autoregulation within 72 hours after inoculation (judged by time delays). From this result we surmise that the fully metabolic wave of autoregulation signal coming from the shoot affected the root perhaps 5 to 7 days after the initial inoculation. We want to stress here that many genetic and experimental parameters alter the extent and timing of autoregulation, such as seedling age, bacterial culture age and dose (Caetano-Anollés and Gresshoff, 1991a).

The shoot suppressive signal was termed SDI (for shoot derived inhibitor). That the signal is an inhibitor in the wild-type and not an activator in the mutant was resolved by an experiment, in which uninoculated and inoculated shoots were grafted after differing periods of being challenged by the inoculation onto uninoculated roots (Gresshoff et al, 1989). In other words, we asked whether a wild-type or nts382 shoot, being attached to an inoculated root system will produce either an inhibitor or an activator. The data suggested that the signal was an inhibitor that was produced in the wild-type plant.

It is possible that the autoregulation signal does not act directly on the meristematic pre-nodule centers. SDI may increase the speed by which a root region matures to limit the growth of pre-nodules below a certain size. Alternatively SDI may slow the development of the pre-nodule, so that the normal developmental transition from the infectable growing root tip region to a non-dividing root tissue inhibits further pre-nodule growth. In the absence of any precise chemical data on the structure and function of the postulated signals, only models designed to help experimental strategies and further searches can be presented.

Caetano-Anollés et al (1991b) resolved the question whether the primordia are arrested or aborted during autoregulation. Surgical nodule removal from wild-type and nts382 plants allowed formation of nodules in the same tissues from where the original nodules were excised. Because nodule redevelopment occurred in both genotypes we postulate an additional control mechanism in which nodules in more advanced stages (i.e. mature and fully emerged nodules) suppress underdeveloped stages. Removal of near-by nodules could signal the plant to release arrested nodule primordia. This developmental strategy appears as a delayed regulatory response. Rather than controlling nodule initiation and infection, soybeans 'permit' the development of some primordia to a sufficiently advanced stage (stage IV), so that they are available as 'spare' nodules, should the original nodule complement be destroyed by predation or drought.

We have attempted to reveal some of the molecular components involved in these regulating circuits. Significant advance has occurred in the genetic and molecular characterization of the nts locus (Landau-Ellis et al, 1991; Funke and Gresshoff, submitted). Biochemical studies have detected developmental polypeptide changes in leaves of mutant and wild-type plants. However, none of the changes were associated with the inoculation or mutant status (Sayavedra-Soto et al, 1991). Either the detection methods (two dimensional electrophoresis of in vitro translation products, total protein or in vivo labelled protein) were insufficient to detect a subtle change, or the involvement of the leaf in autoregulation is constitutive and depends on root-derived signals at the biochemical but not genetic level.

Likewise, we have attempted to characterize the signal molecules stemming from the shoot of wild-type and mutant plants. Although initial data suggested an involvement of abscisic acid (Gresshoff et al, 1988; also see Table 1), these were not confirmed by later studies. Whether bioassays and subsequent chemical fractionation of shoot-derived fractions can reveal the nature of SDI depends on the stability, complexity and multiplicity of factors within such extracts.

At present, it is hoped that the eventual isolation of the coding and regulatory sequences of the nts locus will provide additional means by which the physiological action of nts is understood. The understanding of one part of the systemic control mechanism may provide us with information of related circuits involved in the regulation of the number of other plant parts like lateral roots, flowers and buds.

References

Allen, O.N. & Allen, E.K. (1981) In: The leguminosae, a source Book of characteristics, uses and nodulation. Univ. Wisconsin Press, Madison, WI.

Bauer, W.D. (1981) *Annu. Review of Plant Physiol.* **32**,407-447.

Bender, G.L. & Rolfe, B.G. (1985) *Plant Sci.* **38**,135-140.

Bhuvaneswari, T.V., Turgeon, G.B. & Bauer,W.D. (1980) *Plant Physiol.* **66**,1027-1031.

Bogusz, D., Llewellyn, D., Craig, S., Dennis, E. & Appleby, C. (1990) *Plant Cell* **2**,633-641.

Buzzell, R.I., Buttery, B.R. & Ablett,G.(1990). In : Nitrogen Fixation: Achievements and Objectives. Gresshoff, P.M., Roth, L.E., Stacey, G. & Newton, W.E. (eds.), Chapman and Hall, New York p. 726.

Caetano-Anollés, G. & Bauer, W.D. (1988) *Planta* **175**,546-557.

Caetano-Anollés, G. & Gresshoff, P.M. (1990) *Plant Science* **71**,69-81.

Caetano-Anollés, G. & Gresshoff, P.M. (1991a) *Applied Environmental Microbiol* in press.

Caetano-Anollés, G. & Gresshoff, P.M. (1991b) *J. Plant Physiol.* in press.

Caetano-Anollés, G. & Gresshoff, P.M. (1991c) *Plant Physiol.* **95**,366-373.

Caetano-Anollés, G. & Gresshoff, P.M. (1991d) *Annu Rev. Microbiol.* **45**,345-382.

Caetano-Anollés, G., Paparozzi, E., & Gresshoff, P.M. (1991) *J. Plant Physiol.* **137**,389-396.

Caetano-Anollés, G., Joshi, P.A., & Gresshoff, P.M. (1991) *Planta* **183**,77-82.

Calvert, H. E., Pence, M. K., Pierce, M., Malik, N. S. & Bauer, W. D. (1984) *Can J. Bot.* **62**,2375-2384.

Carroll, B.J. & Gresshoff, P.M. (1983) *Zt. Pflanzenphysiol.* **110**,77-88.

Carroll, B.J., McNeil, D.L., & Gresshoff, P.M. (1985a) *Plant Physiol.* **78**,34-40.

Carroll, B.J., McNeil, D.L., & Gresshoff, P.M. (1985b) *Proc. Natl. Acad. Sci. (USA)* **82**,4162-4166.

Carroll, B.J., McNeil, D.L. & Gresshoff, P.M. (1986) *Plant Sci.* **47**,109-114.

Carroll, B.J. & Mathews, A. (1990). In: Molecular biology of symbiotic nitrogen fixation, Gresshoff, P.M. (ed.), CRC Press, Inc., Boca Raton, FL. pp. 159-180.

Carroll, B.J., Hansen, A.P., McNeil, D.L. & Gresshoff, P.M. (1987) *Aust. J. Plant Physiol.*. **14**,679-687.

Day, D.A., Carroll, B.J., Delves, A. & Gresshoff, P.M. (1989) *Physiol. Plantarum.* **75**,37-42.

Debruijn, F., Hilgert, U., Stigter, J., Schneider, M., Meyer, H., Klosse, U. & Pawlowski. (1990) In: Nitrogen Fixation: Achievements and Objectives, Gresshoff, P.M., Roth, L.E., Stacey, G. & Newton, W.E. (eds.), Chapman and Hall, New York pp. 33-44.

Delves, A.C., Carter, A., Carroll, B.J., Mathews, A., & Gresshoff, P.M. (1986) *Plant Physiol.* **82**,588-590.

Delves, A.C., Higgins, A. & Gresshoff, P.M. (1987) *J. Plant Physiol.* **128**,473-478.

Delves, A. , Higgins, A. & Gresshoff, P.M. (1987) *Aust. J. Plant Physiol.* **14**, 689-694.

Delves, A.C., Higgins, A. & Gresshoff, P.M. (1991) *Plant, Cell, and Environment,* in press.

Diaz, C.L., Melchers, L.S., Hooykaas, P.J., Lugtenberg, B.J., & Kijne, J.W. (1989) *Nature* **338**,579-581.

Dickstein,R., Bisseling, T., Reinhold, V. & Ausubel, F. (1988) Genes & Develop. **2**,677-687.

Duc, G. & Messenger A. (1989) *Plant Sci.* **60**,207-213.

Farmer, E.E. & Ryan, C.A. (1990) *Proc. Natl. Acad. Sci. USA* **87**,7713-7716.

Fortin, M.G., Zelechowska, M. & Verma, D.P.S. (1985) *EMBO J.* **4**,3041-3046.

Fortin, M.G., Morrison, N.A. & Verma, D.P.S. (1987) *Nucleic Acids Res.* **15**,813-824.

Franssen, H.J., Nap, J.P., Gloudemans, T., Stiekema, W., Van Dam, H., et al (1987) *Proc. Natl. Acad. Sci. USA* **84**,4495-4499.

Franssen, H.J., Scheres, B., van der Weil, C., Horvath, B., Moerman, M., Yang, W.C., Govers, F., & Bisseling, T. (1990) In: Nitrogen Fixation: Achievements and Objectives, Gresshoff, P.M., Roth, L.E., Stacey, G. & Newton, W.E. (eds.), Chapman & Hall, Inc., New York, pp. 709-712.

Gerahty, N., Caetano-Anollés, G., Joshi, P.A. & Gresshoff, P.M. (1990) In: Nitrogen Fixation: Achievements and Objectives, Gresshoff, P.M., Roth, L.E., Stacey, G. & Newton, W.E. (eds.), Chapman and Hall, New York, pp. 737.

Gremaud, M.F. & Harper, J.E. (1988) *Plant Physiol.* **89**,169-173.

Gresshoff, P.M., Day, D.A., Delves, A.C., Matthews, A., et al (1985) In: Nitrogen Fixation research progress Evans, H.J., Bottomley, P.J. & Newton, W.E. (eds.), Martinuh Nijhoff, Dordretch, pp. 19-25.

Gresshoff, P.M. & Delves, A.C. (1986) In: Plant Gene Research III, Blonestein, A.D. & P.J. King (eds.), Springer Verlag, Wien, pp. 159-206.

Gresshoff, P.M., Krotzky A.J., Mathews, A., Day, D.A., Schuller, K.A., Olsson, J., Delves, A.C. & Carroll, B.J. (1988) *Plant Physiol.* **132**,417-423.

Gresshoff, P.M., Olsson, J.E., Li, Z.Z. & Caetano-Anollé.G. (1989) *Cur. Topics in Plant Biochem.& Plant Physiol.* **8**,125-139.

Hansen, A.P., Gresshoff, P M., Pate, J.S.& Day, D.A. (1990) *J. Plant Physiol.* **136**,172-179.

Hinson, K. .(1975) *Agronomy J* **67**,799-804.

Hunt, S., King, B.J. & Layzell, D.B. (1989) *Plant Physiol.* **91**,315-321.

Jacobsen , E. (1984) *Plant and Soil* **82**,427-438.

Kosslak, R.M. & Bohlool, B.B.(1984) *Plant Physiol.* **75**,125-130.

Kouchi, H., Tsukamoto, M., & Tajima, S. (1989) *Plant Physiol.* **135**,608-617.

Landau-Ellis, D., Angermüller, S., Shoemaker, R. & Gresshoff, P.M.(1991) *Molec. Gen. Genet.* in press.

Layzell, D., Hunt, S., Moloney, S.M., & Del Castillo, L.D. (1990) In: Nitrogen Fixation: Achievements and Objectives, Gresshoff, P.M., Roth,L.E., Stacey, G. & Newton, W.E. (eds.), Chapman and Hall, New York, pp 21-32.

Lee J.S. &Verma, D.P.S. (1984) *EMBO J.* **3**,2745-2752.

Legocki, R.P. & Verma, D.P.S. (1980) *Cell* **20**,153-163.

Lerouge, P., Roche, P., Faucher, C., Maillet, F., & Truchet, F. (1990) *Nature* **244**,781-784.

Letham, D.S., Higgins, T.J., & Goodwin , P. (1978) In: Phytohormones and Related Compounds - A Comprehensive Treatise. Vol I and II, Elsevier/North Holland Press.

Long, S.R., Fisher, R.F., Ogawa, J., Swanson, J., Ehrhardt, E.M.,Atkinson, E.M. & Schwedock, J.S. (1991) In: Advances in molecular genetics of plant-microbe interactions, Hennecke, H. and Verma, D.P.S. (eds.), Kluwer Academic Publishers, The Netherlands, pp. 127-133.

Mathews, A., Carroll, B.J. & Gresshoff, P.M. (1989a) *Protoplasma* **150**,40-47.

Mathews, A., Carroll, B.J. & Gresshoff, P.M.(1989b) *J. Heredity* **80**,357-360.

Mellor, R. & Werner, D.(1990) In: Molecular Biology of Symbiotic Nitrogen Fixation. Gresshoff, P.M. (ed.), CRC Press, Inc., Boca Raton, FL., pp. 111-130.

Nap, J.P. & Bisseling, T. (1990) In: Molecular Biology of Symbiotic Nitrogen Fixation, Gresshoff, P. M. (ed.), CRC Press, Inc., Boca Raton, FL., pp. 181-229.

Nguyen T., Zelechowska, M.G., Foster, V., Bergmann, H. & Verman, D.P.S. (1985) *Proc. Natl. Acad. Sci.* **82**,5040-5044.

Norris, J.H., Macol, L.A. & Hirsch, A.M.(1988) *Plant Physiol.* **88**,321-28.

Nutman, P.S. (1952) *Ann. Bot. N. S.***16**,81-102.

Olsson, J.E., Ph D. (1988) Dissertation. Australian National University, Canberra.

Olsson, J., Nakao, P., Bohlool, B.B. & Gresshoff, P.M. (1989) *Plant Physiol.* **90**,1347-1352.

Pierce, M., & Bauer, W.D. (1983) *Plant Physiol.* **73**,286-290.

Price, G.D., Mohapatra, S.S. & Gresshoff, P.M. (1984) *Bot. Gaz.* **145**.444-451.

Ridge, R. & Rolfe, B. G.(1985) *J. Plant Physiol.* **122**,121-137.

Sanchez, F., Padilla, J.E., Pérez, H. & Lara M. (1991) *Annu. Rev. Plant Physiol. & Plant Mol. Biol.* **42**,507-528.

Sargent, L.,Huang, S.Z., Rolfe, B.G. & Djordjevic, M.A. (1987) *Applied & Environm. Microbiol.* **53**,1611-1619.

Sayavedra-Soto, L.A., Angermüller, S.A., Prabhu, R. & Gresshoff, P.M. (1991) *J. Plant Physiol* in review.

Scott, K. & Bender, G. (1990) In: Molecular Biology of Symbiotic Nitrogen Fixation. Gresshoff, P.M. (ed.), CRC Press, Inc., Boca Raton, FL., pp. 231-252.

Sprent, J.I. (1990) In: Nitrogen Fixation: Achievements and Objectives, Gresshoff, P.M., Roth, L.E., Stacey, G. & Newton,W.E. (eds.), Chapman and Hall, New York, pp. 45-54.

Suzuki, H. & Verma, D.P.S. (1991) *Plant Physiol.* **95**,384-389.

Szabados, L., Ratet, P., Grunenberg, B. & de Bruijn, F.(1990) *J. Plant Cell* **2**,973-986.

Trese, A.T. & Pueppke, S.G. (1990) *Plant Physiol.* **92**,946-953.

Thummler and Verma, D.P.S. (1987) *J. Biol. Chem.* **262**,14730-14736.

Truchet, G., Roche, P., Lerouge, P., Vasse, J., Camut, S.M, de Billy, F., Promé. J.C. & Dénarié, (1991) *J. Nature* **351**,670-673.

van Kessel C & Singleton, P. W. (1987) *Plant Physiol.* **83**,552-556.

Verma, D.P.S. (1990). In: Nitrogen Fixation: Achievements and Objectives, Gresshoff, P.M., Roth, L.E., Stacey, G. & Newton, W.E. (eds.), Chapman and Hall, New York pp. 235-237.

Vincent, J. M. (1980) In: Nitrogen Fixation, Vol. 2, Newton, W. E. & Orme-Johnson, W. H. (eds.), Univ. Park. Press, Baltimore, pp. 103-129.

Weeden, N.F., Kneen, B.E. & LaRue, T.A. (1990). In: Nitrogen Fixation: Achievements and Objectives, Gresshoff, P.M., Roth, L.E., Stacey, G. & Newton, W.E. (eds.), Chapman and Hall, New York pp. 323-330.

RFLP Linkage analysis of symbiotic mutants of soybean

Deborah Landau-Ellis, Randy Shoemaker*, Sieglinde
Angermüller, and Peter M. Gresshoff

*Plant Molecular Genetics and Center for Legume Research, The University of
Tennessee, Knoxville, TN 37901-1071, USA;*

*USDA/ARS,Iowa State University, Ames, IA 50011, USA

Introduction

The objective of this study is to determine linkage of soybean (*Glycine max* (L.)
Merr.) nodulation phenotypes to RFLP markers in order to localize genes associated
with nodulation. Soybean nodulation mutants, nts382 and nts1007
(supernodulators), and nod49 and nod139 (nonnodulators) were originally isolated
from M2 and M3 populations of ethyl methane sulfonate mutagenized seeds of a *G.
max* cultivar Bragg (Carroll et al., 1985a,b). This common origin resulted in a high
level of molecular homology among Bragg and the four mutants. A more diverse
population was needed in order to generate segregation and detect linkage
(cosegregation) to an RFLP marker (Keim et al., 1989).

The genus *Glycine* subgenus *Soja* consists of two species, the cultivated soybean
(*Glycine max* (L.) Merr.) and the wild annual soybean (*Glycine soja* Sieb. and Zucc.).
In order to obtain a more diverse population, crosses were made between Bragg, the
four mutants, and a *G. soja* plant introduction (PI468.397). Linkage analysis is being
conducted on each of the segregating F2 populations resulting from these crosses.

Iowa State University has supplied approximately 200 plasmid clones which detect
polymorphisms between their experimental line A81-356022 (*G. max*) and PI468.916
(*G. soja*). These probes make up a large portion of a linkage map which currently
covers 30 linkage groups, 293 loci, and 2400 centimorgans (linkage assignments were
based on segregation analysis of 60 F2 plants from the cross A81-356022 X PI468.916).
The genomic size of soybean has been represented as 1.07 X 10^9 bp (Blackhall et al.,
1991) and as 1.29 X 10^9 bp (Gurley et al, 1979) suggesting 3.5 - 4.5 Mb DNA per locus or
approximately 500 kb cM^{-1}. This average value may be as low as 250 kb cM^{-1} as the
RFLP map does not include the low recombinogenic heterochromatin, which
comprises about 50% of the soybean genome (Singh and Hymowitz, 1988)

In general, functionally related genes do not cluster on a specific chromosome but are distributed across the genome (Suzuki et al., 1981). In garden pea (*Pisum sativum* L.), nodulation related genes have been found on 6 of the 7 chromosomes; however there is a clustering of *sym* genes on chromosome 1 (Weeden et al, 1990). In our soybean genotypes complementation analysis shows that *nts382* and *nts1007* share the same complementation group (Delves et al, 1988) while *nod49* and *nod139* are on separate independently segregating complementation groups from the nts allele as well as from each other (Mathews et al., 1989, 1990). The nonnodulator *nod49* is complementary to the naturally occuring *rj1* allele.

Plant material and crossing

The genus *Glycine* subgenus *Soja* consists of two species. The cultivated soybean (*Glycine max* (L.) Merr.) has an erect and bushy growth habit, and the wild annual soybean (*Glycine soja* Sieb. and Zucc.) has a vining or trailing growth habit. Both species have the same chromosome number 2n=40 and readily hybridize (Singh and Hymowitz, 1988). In this study *Glycine max* (L) Merrill cv Bragg and each of the four mutants were crossed to *G. soja* PI468.397 (kindly supplied to us by Dr. Gary Stacey, CLR, University of Tennessee, Knoxville) (see Table 1).

Table 1. Plant material used in our crossing strategy

Wildtype nodulation	Symbiotic mutants of Bragg	
Glycine max cultivar Bragg ♀	nts1007	supernodulating ♀
	nts382	supernodulating ♀
Glycine soja PI468.397 ♂	nod139	nonnodulating ♀
	nod49	nonnodulating ♀

Crossing was done under field conditions at the University of Tennessee's Knoxville Experiment Station (see Table 2). In each case *G. soja* (purple flower) was used as the male parent while *G. max* (white flower) was used as the female parent (expressing the homozygous recessive mutant trait). These crosses gave frequencies of success between 1 and 11% which reflected experimental and environmental conditions at the time of crossing.

Probes

The USDA-ARS/ISU soybean linkage map consists of almost 300 well spaced markers which were derived from a *Pst*I library of an experimental line *G. max* A81-356022. About 30% of the clones detected polymorphisms in their parental material, *G. max* line A81-356022 and *G. soja* PI468.916 (Keim et al, 1990). Some of these probes

(9 out of 28) have not detected polymorphisms between the Bragg genotypes and PI468.397. From the probes used thus far which do exhibit polymorphisms between Bragg and PI468.397, none show detectable differences among Bragg, nts1007, nts382, nod139, and nod49, emphasizing the common origin of these (near isogenic) symbiotic mutants.

Table 2. Original crosses from the summer of 1988. Percent success rate is based on the number of crosses attempted and the number of resulting pods.

Parental lines used for cross	Number of crosses	Number of pods	Number of seeds	Success rate (%)
♀ X ♂				
nts1007 X PI468.397	83	6	8	7.2
nts382 X PI468.397	177	13	14	7.3
nod139 X PI468.397	78	1	1	1.3
nod49 X PI468.397	90	10	14	11.1
TOTAL	433	30	37	6.9

Southern Hybridization

Purified DNA was restricted with appropriate endonucleases (known to exhibit polymorphisms), and separated by agarose gel electrophoresis (20μg DNA/lane, 0.9% agarose, 1X TAE electrophoresis buffer, 35V, 15h). The DNA from the gel was then transferred onto Zeta-Probe® Nylon membrane (Bio-Rad) via vacuum blotting (LKB VacuGene). Membranes were probed with radioactively (αdCT^{32}P, random primer method, Boehringer Mannheim) labelled inserts from clones known to be polymorphic between the two parents. Hybridizations were performed at 60°C according to Zeta-Probe instructions (Bio-Rad). Membranes were exposed to Kodak X-Omat AR film from 24 to 72 hours at -70°C.

F1 Hybrid plants

F1 progeny were grown in the greenhouse on artificial growth media and were inoculated with *Bradyrhizobium japonicum* strain USDA110. These plants were confirmed as F1 hybrids by their root phenotype (wild type), purple flower color (purple is dominant to white), intermediate growth habit, seed size and color, as well as their hybrid RFLP patterns (Figure 1).

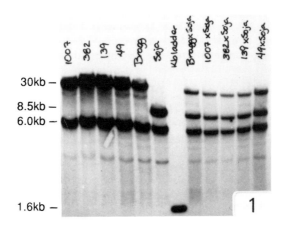

Figure 1. Autoradiograph of EcoRI digested DNA from parental (the first 6 lanes) as well as F1 hybrid plants (the last 5 lanes) probed with pK-19. The F1 plants inherited half of their DNA from each parent; therefore F1 banding patterns contain the 30kb band of the G. max parents, the 8.5kb band of the G. soja parent (PI468.397), and the 6.0kb fragment common to both parents. Notice that there are no RFLPs among Bragg and the 4 symbiotic mutants.

Linkage analysis and F2 segregation

Linkage between genetic markers refers to the predominance of parental combinations among the offspring of crosses. Linked genetic markers tend to cosegregate rather than recombine at random during meiosis due to cross over events. The absence of linkage between two markers results in independent assortment when the markers are distal to each other (50 centimorgans or more) on the same chromosome or when they are located on separate chromosomes.

A sub-population (82 plants) of F2 progeny seed, from a confirmed F1 plant (designated as C-16) of the cross *G. max* line nts382 X *G. soja* PI468.397, was grown on artificial soil media (Turface) and inoculated with *B. japonicum* USDA110. These plants were screened after 28 days for nodulation phenotype. There were 20 *nts* segregants from this population, exhibiting the expected ratio, of 3 wild type : 1 supernodulator, for a single Mendelian locus. These 20 supernodulating F2 plants (homozygous recessive) were repotted and grown vegetatively in order to collect young leaves for DNA isolation using methods described elsewhere (Dellaporta et al, 1983).

These 20 F2 plants, although homozygous for their nodulation genotype, were segregating for many other traits including RFLP banding patterns. The strategy then required the detection of cosegregation of the *nts* locus with an RFLP marker. We initially selected 1 to 3 probes from each of the linkage groups in order span the genome. Because not all the probes utilized from the USDA-ARS/ISU soybean linkage bank showed RFLPs for the genetic lines used in this study, mapped markers were first confirmed to be polymorphic between *G. max* cv Bragg and *G. soja* PI468.397 before being tested for segregation on the F2 populations. Chi square (χ^2) goodness of fit tests were used to determine the probability that the ratios of segregation in the F2 plant material fit the expected ratios for given recombination frequencies.

The clone pK-19 from linkage group J was used to probe 17 of the 20 nts segregants from this F2 population (Figure 2). Chi-square analysis indicates that this segregation is not different than a 1:2:1 ratio of independent assortment or 50% recombination (Table 3).

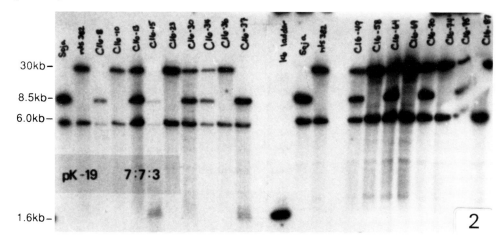

Figure 2. The parentals as well as seventeen of the 20 nts segregants were restricted with EcoRI and hybridized with clone pK-19. G. soja = PI468.397, nts382 =G. max cv Bragg (line nts382). F2 plants are designated as C16-x to allow comparisons with other autoradiographs. The F2 ratio was 7:7:3 for G. max (nts), heterozygous, and G. soja banding patterns respectively.

Table 3. Chi-square goodness of fit test for probe pK-19 for independent assortment.

H_0: Segregation is not different from a 1 : 2 : 1 ratio of 50% recombination.
H_1: Segregation is different from a 1 : 2 : 1 ratio of 50% recombination.

Class (banding pattern)	Observed (O)	Expected (E)	$(O-E)^2$	$(O-E)^2/E$
G. *max* (nts382)	7	4.25	7.56	1.78
Heterozygous	7	8.50	2.25	0.26
G. *soja*	3	4.25	1.56	0.37
	17	17.00		χ^2=2.41

d.f. = 2 α = 0.05 χ^2 calculated (c) = 2.41 χ^2 tabular (t) = 5.99
$\chi^2{}_c < \chi^2{}_t$ ∴ Accept H_0; conclude that, at a = 0.05, segregation is not different from 50% recombination. The probe pK-19 segregates independently from the *nts* locus.

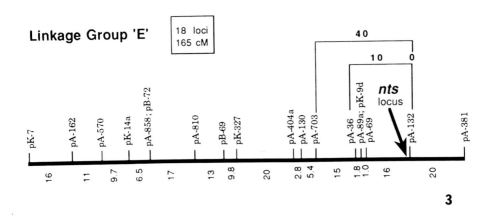

Figure 3. Linkage Group 'E' from the USDA-ARS/ISU soybean linkage map.

Linkage to the *nts* supernodulating locus

Our results show that several probes on tentative linkage group 'E' cosegregate with the nts supernodulation phenotype/genotype (Figure 3). This linkage group will eventually correspond to one of the 20 chromosomes in soybean. The RFLP banding patterns for probe pA-36 segregate 15:4:0 for *G. max* (nts), heterozygous, and *G. soja* respectively (Figure 4). The chi-square goodness of fit test shows that this ratio fits the pattern for 10% recombination (Table 4). This locates the probe pA-36 approximately 10 cM distal to the *nts* locus. Probe pA-703 (on the same linkage group, autoradiograph not shown) also cosegregates with the *nts* locus at a distance of ~40 cM; χ^2 analysis shows that the ratio 6:8:3 is not different than 1:2:1 or independent assortment.

Table 4. Chi-square goodness of fit test for probe pA-36 at 10cM distal to nts representing 10% recombination.

H_0: Segregation is not different from a 15.36 : 3.61 : 0.19 ratio of 10% recombination.
H_1: Segregation is different from a 15.36 : 3.61 : 0.19 ratio of 10% recombination.

Class (banding pattern)	Observed (O)	Expected (E)	$(O-E)^2$	$(O-E)^2/E$
G. max (nts382)	15	15.36	0.1296	0.0084
Heterozygous	4	3.61	0.1521	0.0421
G. soja	0	0.19	0.0361	0.19
	19	19.00		$\chi^2 = 0.24$

d.f. = 2 $\alpha = 0.05$ χ^2 calculated (c) = 0.24 χ^2 tabular (t) = 5.99
$\chi^2_c < \chi^2_t$ ∴ Accept H_0; conclude that, at a = 0.05, segregation is not different from 10% recombination. The probe pA-36 is ~ 10 map units from the *nts* locus.

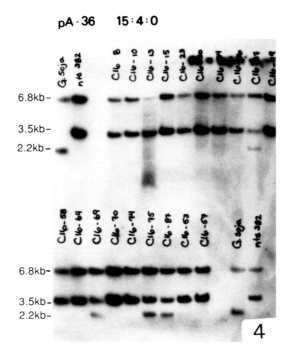

Figure 4. Southern Hybridization of probe pA-36 to DNA of 19 HindIII restricted super-nodulating segregants, reveals predominantly nts382 banding patterns suggesting cosegregation with the nts locus. Parental as well as F2 DNA is represented. G. soja=PI468.397, nts382 =G. max cv Bragg (line nts382), F2 plants are designated as C16-x. Banding patterns of F2 plants segregate 15:4:0. Fifteen of these F2 plants have the 3.5kb band of the nts parent plus the 6.8kb band common to both parents. Four of the lanes show the hybrid banding pattern with the 3.5kb band of the nts parent, the 2.2kb band of the G. soja parent, as well as the common 6.8kb band. No G. soja banding patterns are present in these 19 supernodulating F2 plants.

Probe pUTG-132a is tightly linked to the supernodulating (nts) locus

The probe pA-132 is located on the other side of pA-36 from pA-703 (linkage group E). When DNA from these same supernodulating F2 plants was restricted with DraI and probed with pA-132, tight linkage was observed. Only banding patterns of the nts382 parent were observed in these segregants (Figure 5) suggesting a lack of recombination in this region between the nts locus and this clone. The probe pA-132 exists as 3 separate PstI fragments and gives a complicated banding pattern (not shown). The uppermost of these fragments (1.8kb) shows the linked polymorphism and is designated as pUTG-132a.

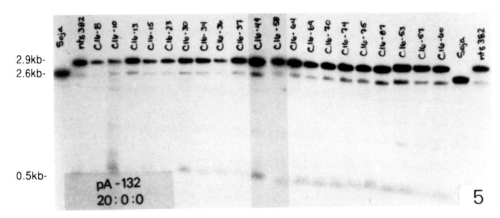

2.9kb-
2.6kb-

0.5kb-

Figure 5. *Southern Hybridization of parental and F2 generation supernodulating segregants. G. soja=PI468.397, nts382=G. max cv Bragg (line nts382), F2 plants are designated as C16-x to allow comparisons of individual plants with their counterparts in other autoradiographs. DNA samples have been restricted with the endonuclease DraI and probed with pUTG-132a. Every F2 plant exhibits the banding pattern of the nts382 parent, the intense 2.9kb band as well as the less intense 2.6kb band. None of these 20 F2 plants show the single intense band of the G. soja parent nor of the hybrid banding pattern which would contain the 2.9kb band plus a more intense 2.6kb band.*

Discussion

Results were obtained from 20 supernodulating segregants out of an 82 plant F2 subpopulation of the *G. soja* X nts382 cross (C-16). A larger population will give a more accurate indication of recombination distances. In this study we only looked at the supernodulating segregants and did not consider RFLP segregation ratios for the rest of the population. Analysis is presently in progress on an additional F2 subpopulaton of C-16 plants. This will include, in addition to the supernodulating plants, the plants homozygous dominant for wild-type nodulation, and plants heterozygous for nodulation genotype. It is expected that these probes will segregate independently across the total population but we want to see if there is any skewing towards *G. max* or *G. soja* patterns.

Recombination frequencies on linkage group 'E', obtained from our genetic material, differ from those represented on the USDA-ARS/ISU soybean linkage map (Figure 3). These discrepancies between recombination distances could be the result of error due to small population size. The strength of linkage between markers, or recombination frequency, may only be considered constant in experiments carried out under the same conditions; the recombination process may also be sensitive to environmental conditions. In addition to using different genetic material, our F1 hybrid plants were grown at another location and time which most likely offered different environmental conditions at the time of meiosis. The fact that we only screened the supernodulating segment of the population may also account for our incongruent results.

Similar screening is in progress for the other symbiotic soybean mutants. It is expected that F2 plants from the cross of nts1007 X PI468.397 will give the same results and recombination frequencies obtained from the cross nts382 X PI468.397. The two supernodulators nts382 and nts1007 share the same complementation group. It will be interesting to find the loci controlling nodulation phenotype for the two nonnodulators nod139 and nod49 which are on separate complementation groups.

Acknowledgement

The research was supported by an endowment to the Racheff Chair of Excellence, USDA CSRS grant no. TEN00879, Pioneer Hi-Bred International, and the Tennessee Soybean Promotion Board. Ms. Marcia Young and Lesley Schuller provided technical assistance. The figures 1-5 were used with the permission of Springer-Verlag.

References

Blackhall, N. W., Hammatt, N., and Davey, M. R. (1991) Soybean Genetics Newsletter (in press).

Carroll, B. J., McNeil, D. L., and Gresshoff, P. M. (1985a) Plant Physiology 78, 34-40.

Carroll, B. J., McNeil, D. L., and Gresshoff, P. M. (1985b) Proceedings of the National Academy of Science 82, 4162-4166.

Dellaporta, S.L. Wood J., and Hicks, J. B. (1983) Plant Molecular Biology Reporter 1, 19-21.

Delves, A. C., Carroll, B. J., and Gresshoff, P. M. (1988) Journal of Genetics 67, 1-8.

Gurley, W. B., Hepburn, A. G., and Key, J. L. (1979) Biochimica et Biophysica Acta 561, 167-183.

Keim, P., Shoemaker, R. C., and Palmer, R. G. (1989) Theoretical and Applied Genetics 77, 786-792.

Mathews, A., Carroll, B. J., and Gresshoff, P. M. (1989) The Journal of Heredity 80, 257-360.

Mathews, A., Carroll, B. J., and Gresshoff, P. M. (1990) Theoretical and Applied Genetics 79, 125-130.

Singh, R. J., and Hymowitz, T. (1988) Theoretical and Applied Genetics 76, 705-711.

Suzuki, D. T., Griffiths, A. J. F., and Lewontin, R. C. (1981) An introduction to genetic analysis. W. H. Freeman and Company. New York, San Francisco.

Weeden, N. F., Kneen, B. and LaRue, T. (1990) In: Nitrogen Fixation: Achievements and Objectives, Gresshoff, P. M., Roth, L. E., Stacey, G., and Newton. W. E. (eds.), Chapman and Hall, New York.

Physical mapping of the *nts* region of the soybean genome using pulse field gel electrophoresis (PFGE)

Roel P. Funke and Peter M. Gresshoff

Plant Molecular Genetics and Center of Legume Research, The University of Tennessee, Knoxville, TN 37901-1071, USA.

Introduction

Pulse field gel electrophoresis (PFGE) is a technique for resolving large fragments of DNA ranging in size from 50 kb up to 12 Mb or more (Orbach et al, 1988). It is being used to generate electrophoretic karyotypes of simple eukaryotes (Mills and McCluskey, 1990) and long range restriction maps of mammalian genomes (Petit et al, 1990). More recently the technique has been adapted for use in higher plants, including tomato (Ganal et al, 1990), barley and rye (Cheung et al, 1990), rice (Sobral et al, 1990), and soybean (Funke et al, 1990; Honeycutt et al, 1990).

PFGE will see increased use in the study of higher plant genomes. In addition to physical mapping regions of particular chromosomes, the technique can be used to relate genetic distance to physical distance in these regions. This relationship varies throughout the genome (Rick, 1971), and PFGE can be used to identify areas of unusually high or low recombinational activity. Ganal et al (1989) used this approach to demonstrate that in the *Tm-2a* region of the tomato genome there is an approximately sevenfold reduction of recombinational activity over an average expected value calculated from the number of centiMorgans (cM) on the genetic map of tomato and the estimated genome size in bp. In addition, the relationship between physical and genetic distance is a critical yardstick to possess en route to the cloning of genes that have no defined biochemical product. This information gives an indication of the physical distance that must be covered in search of the gene starting from a polymorphic marker that has been found to be linked to it. Finally, a chromosome walk in search of candidate genes would be greatly accelerated by the use of YACs (yeast artificial chromosomes, 100-1000 kb segments of cloned DNA transmitted in yeast) rather than conventional λ or cosmid clones (maximum size about 20 kb), and YACs are currently best characterized by PFGE.

In this chapter we outline a procedure for the isolation of intact chromosomal DNA from soybean, its restriction with rare-cutting endonucleases, and Southern blotting of large fragments of DNA resolved by PFGE. In addition, we will relate physical and recombinational distances in a cluster of polymorphic markers near the *nts* gene, which has been mapped to within 1 cM of the polymorphic marker pUTG132a in linkage group E of the USDA-ARS/ISU RFLP map of soybean (Landau-Ellis et al, in press).

Procedures

Protoplasts

Protoplasts were obtained from fresh small (1-2cm) leaves taken from greenhouse grown wild type plants. Leaves were surface-sterilized and cut into 1-2 mm ribbons and transferred to a digestion mixture consisting of 2% driselase (Sigma Chemicals), 1% cellulase Onozuka R-10 (Serva), and 0.2% pectolyase Y-10 (Sigma Chemicals). The enzymes were dissolved in a buffer containing 36.4 g mannitol, 735 mg $CaCl_2.7H_2O$, and 487 mg MES per 500 ml. A 10 ml digestion mixture was placed on a slowly rotating shaker for 5 hours.

Following digestion, the mixture was sieved consecutively through 80 and 43 µm nylon filters. Protoplasts were collected by centrifugation at 100xg for 10 minutes. The cell concentration was adjusted to about $4x10^7$ per ml and the suspension was mixed with an equal volume of 1% low gelling temperature agarose (FMC INCERT) at 42 °C. 200 µl aliquots of the protoplast-agarose mixture were pipetted into a pre-chilled mold and allowed to solidify before proceeding to lysis of the embedded cells.

Cell lysis

Plugs were transferred to at least three times their combined volume of ESP (0.5M EDTA, 1% sarkosyl, 1 mg per ml proteinase K) at 50 °C while being gently agitated at 50 rpm. After 24 hours the solution was replaced with fresh ESP and left for a further 24 hours, at which point the plugs could be prepared for restriction digestion.

Restriction of chromosomal DNA in agar plugs

Proteinase K was inactivated by washing the plugs twice for one-half hour in TE buffer containing 1 mM PMSF (100 µl of 100 mM freshly prepared isopropanol stock per 10 ml of TE) at 50 °C. The plugs were rinsed three more times with fresh changes of TE without PMSF.

One-third to one-half of a 200 µl plug (about 5 µg of DNA) was added to 100 µl of 1X restriction enzyme buffer containing 5 mM spermidine and placed on an ice-slush bath for one half hour. Restriction enzyme was added and the mixture was left at 4 °C for an additional half hour to allow the enzyme to diffuse into the plugs before becoming fully active. Digestions were carried out at the recommended temperature and terminated after at least 6 hours by adding 500 µl of ESP and incubating at 50 °C for one hour before commencing electrophoresis.

Electrophoresis and size standards

A CHEF (Bio-Rad) electrophoresis unit was used in all experiments. Running buffer (0.5xTBE) was recirculated through the electrophoresis chamber and maintained at about 14 °C in all experiments. Running conditions varied according to the size range of DNA fragments to be resolved.

For size standards in the 50-600 kb range, oligomers of λ DNA were suitable. These were prepared by allowing λ DNA mixed with LMP agarose to 30 µg per ml to concatemerize at 4 °C. Higher molecular weight size standards were produced by embedding spheroplasts of yeast (*Saccharomyces cerevisea*) in low gelling temperature agarose to a concentration of $1x10^{10}$ cells per ml, and preparing plugs for electrophoresis as described above for soybean protoplasts. Whole chromosomes of yeast are resolved ranging in size from 245 to 2200 kb.

Southern blotting and hybridization

After ethidium bromide staining and destaining of gels for photography, the DNA was depurinated in 0.25M HCl for 10 minutes. DNA was vacuum blotted (LKB) onto Nylon membranes using 0.4 M NaOH under a pressure of 40 cm H_2O for one hour, after which the membranes were neutralized in 2xSSC. DNA was cross-linked to the membranes by treatment with short wave-length UV light using a total energy intensity of 1200 µJ per cm^2.

Probes were labelled to a high specific activity using the random hexanucleotide primer method. Hybridization and washes were carried out at 60 °C and 50 °C, respectively, according to procedures supplied with ZetaProbe membranes (Bio-Rad). Membranes were exposed to XOMAT film overnight at -70 °C.

Results

Fragment Sizes
The gel in Fig. 1 was run with a 5-20 second pulse time for 20 hours at 200 volts. Under these conditions fragments from 10-500 kb are resolved. The autoradiograph demonstrates that with these restriction enzymes the majority of fragments identified by our probes (markers from the soybean RFLP linkage map, USDA-ARS/ISU, supplied by R. Shoemaker, ISU) range in size from 50 to 100 kb. Based on a calculation of the density of markers on the RFLP map, it was expected that in order to compare physical and genetic distances on the map much larger fragments would be needed.

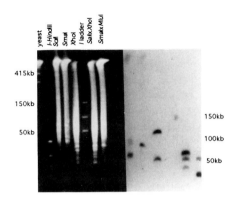

Figure 1. Pulse field gel of soybean DNA and hybridization to the RFLP probe pA-69. Intact chromosomal DNA was restricted with the enzymes shown and electro-phoresed as described in the text. The gel and autoradiograph are not to the same scale.

Comparing physical and genetic distances on the soybean RFLP map

There are estimated to be 1.29×10^9 bp in the haploid genome of soybean (Gurley et al, 1979) and currently the RFLP map spans 2465.69 cM (Keim et al, 1990). Therefore on average there are about 500 kb of DNA per cM on the map. The 293 polymorphic markers on the map are distributed over 26 linkage groups and are separated by an average of 8.4 cM and 4.4 Mb. However, clusters of closely spaced markers occur, and we have focused on such a cluster in linkage group E. This cluster is near the marker pA-132, which has been found to be tightly linked to the *nts* mutant phenotype of soybean (Landau-Ellis et al, in press). Thus for the purpose of starting a systematic search for the *nts* locus, it is of special interest to determine a correlation between genetic and physical distance in this linkage group.

Two markers in this cluster, pK-9 and pA-89, are inseparable by segregation analysis. pA-36 is 1.8 cM from this pair, while pA-69 is 1 cM distant on the other side.

Over the whole soybean genome, there are calculated to be about 500 kb of DNA per cM of genetic distance. Therefore to establish a yardstick of genetic versus physical distance in linkage group E, it would be necessary to resolve fragments at least 500-1000 kb in size. If two polymorphic markers can be shown to lie on the same fragment of DNA, one can assume that they are separated by no more than the number of kb in that fragment. To be confident that two markers do in fact comigrate on the same piece of DNA, the results should be based on a reprobing of the same membrane, and should be reproducible under different conditions of electrophoresis. The conclusion would also be strengthened if comigration could be demonstrated using two different restriction enzymes.

To obtain large fragments of DNA we used restriction enzymes that identify 8 bp sequences, *Not*I and *Sfi*I. The marker pA-69 hybridized to a *Not*I fragment larger than 2.2 Mb in size, while *Not*I and *Sfi*I fragments as small as 400 kb contained pK-9 and pA-89. Three of the markers, pA-36, pK-9 and pA-89, comigrated on the same 900 kb *Not*I fragment, and pK-9 and pA-89 in addition shared 400 kb *Not*I and 500 and 600 kb *Sfi*I fragments. These results are based on reprobings of the same membrane, and they hold true under different conditions of electrophoresis. In separate experiments, we ensured that these markers really are distinct and do not hybridize to one-another.

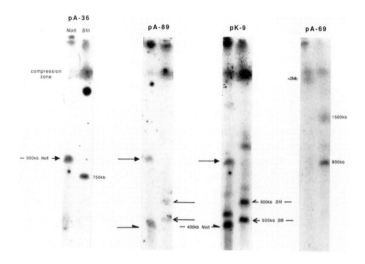

Figure 2. Southern blots of NotI and SfiI digests of soybean DNA. The membrane was probed successively with the four markers shown. Comigration of markers on the same restriction fragments, as discussed in the text, is reproducible under different electrophoresis conditions.

To obtain large fragments of DNA we used restriction enzymes that identify 8 bp sequences, *Not*I and *Sfi*I. The marker pA-69 hybridized to a *Not*I fragment larger than 2.2 Mb in size, while *Not*I and *Sfi*I fragments as small as 400 kb contained pK-9 and pA-89. Three of the markers, pA-36, pK-9 and pA-89, comigrated on the same 900 kb *Not*I fragment, and pK-9 and pA-89 in addition shared 400 kb *Not*I and 500 and 600 kb *Sfi*I fragments. These results are based on reprobings of the same membrane, and they hold true under different conditions of electrophoresis. In separate experiments, we ensured that these markers really are distinct and do not hybridize to one-another.

Discussion

The presence in the genome of multiple copies of some of these polymorphic markers can complicate the interpretation of data like these. In particular, the question arises whether a band on the Southern blot originates from the linkage group E, or from another linkage group which contains a second copy of the marker. Fortunately pA-36 appears to be represented only once in the genome. Thus the 900 kb *Not*I fragment it hybridizes to may be presumed to be from linkage group E. While pK-9 and pA-89 appear on other linkage groups on the RFLP map, the presence of at least one copy of each of these markers on the same 900 kb fragment argues that this fragment is in fact colinear with the pA-36 cluster in linkage group E.

pA-89 has two copies in the genome (it has no internal sites for *Not*I or *Sfi*I), the second one appearing on linkage group I of the RFLP map. While pK-9 has at least three copies in the genome, only one is placed on the RFLP map of soybean, possibly because the others did not show a polymorphism between the parental plants used to construct the map. However, its coincidence with pA-89 on the same 400 kb *Not*I and 500 and 600 kb *Sfi*I fragments suggests that in another linkage group (for example, I) it may also be close to pA-89, so that the relative proximity of these markers is preserved in different parts of the genome.

In summary, these results indicate that in this cluster of markers, 1.8 cM of recombinational distance (the distance between pA-36 and the pair pK-9 and pA-89) is contained by no more than 900 kb of DNA. This agrees with the theoretical value of 500 kb per cM calculated from estimates of the size of the genome in bp and the number of cM currently on the soybean RFLP map. Since the RFLP map does not include low recombinogenic heterochromatin, which comprises 50% of the soybean genome (Singh and Hymowitz, 1988), this value may be revised downward on fine mapping of this region and is likely to vary from one location to another.

The availability of YACs has made it possible to map large fragments of DNA more unequivocally. Using only PFGE it is still possible that different fragments of the same size coincide and comigrate under different conditions of electrophoresis.True confirmation of the findings presented here would come from the construction of a 'contig' of YACs spanning the 900 kb region that appears to harbor the three markers pA-36, pK-9, and pA-89.

References

Cheung, W.Y., & Gale, M.D. (1990) Plant Mol. Biol. **14**, 881-888

Funke, R.P., Sayavedra-Soto, L., & Gresshoff, P.M. (1990) *III Biennal Conference on Molecular and Cellular Biology of Soybean*, July, Iowa State University, Ames

Ganal, M.W., Young, N.D. & Tanksley, S.D. (1989) Mol. Gen. Genet. **215**, 395-400

Gurley, W.B., Hepburn, A.G., & Key, J.L. (1979) Biochim. Biophys. Acta **561**, 167-183

Honeycutt, R.J., Sobral, B.W.S., McClelland, M., & Atherly, A.G. (1990) *III Biennal Conference on Molecular and Cellular Biology of Soybean*, July, Iowa State University, Ames

Landau-Ellis, D., Shoemaker, R., Angermuller, S., & Gresshoff, P.M. (1991) Mol. Gen. Genet. (in press)

Mills, D. & McCluskey, K. (1990) Molecular Plant Microbe Interactions 3, 351-357

Orbach, M.J., Vollrath, D., Davis, R.W. & Yanofsky, C. (1988) Mol.Cell. Biol. **8**, 1469-1473

Petit, C., Levilliers, J., & Weissenbach, J. (1990) Proc. Nat. Acad. Sci. USA **87**, 3680-3684

Rick, C.M. (1971) Stadler Genetic Symp. **1**, 153-174

Sobral, B.W.S., Honeycutt, R.J., Atherly, A.G., & McClelland, M. (1991) Plant Mol. Biol. Rep.8, 253-275

Singh, R.J. & Hymowitz, T. (1988) Theor. Appl. Genet. **76**, 705-711

Gene transfer to barley

R.R. Mendel[1], E. Clauss[2], J. Schulze[2], H.H. Steinbiß[3],
and A. Nerlich[1]

[1] *Institute of Genetics & Crop Plant Research, O - 4325 Gatersleben, Germany*
[2] *Institute of Breeding Research O - 4300 Quedlinburg, Germany*
[3] *Max-Planck-Institute of Breeding Research, W - 5000 Köln 30, Germany*

Introduction

The stable transformation of crop plants belonging to the Gramineae family is of
great importance for all further genetic manipulation. Since agrobacteria do not
work (efficiently) with cereals the techniques of direct DNA transfer have been
developed as an alternative (Klein et al., 1989). We tried different approaches of
directly introducing DNA into cells and organs of barley in order to find a method
for the stable integration of a foreign gene into the genome of barley with the aim to
produce transgenic cell lines as well as transgenic fertile plants that inherit the new
gene to their progeny.

The results of two cell genetic approaches (DNA transfer into protoplasts; DNA
transfer into suspension culture cells by the particle gun) and of two plant genetic
approaches (macroinjection into floral tillers; DNA transfer into growing pollen
tubes) are presented and discussed as preliminary data (part of this work has been
presented previously [Mendel et al., 1990]). A further detailed analysis of the
obtained putative transformants is in progress.

Macroinjection into floral tillers

We used the technique of injecting DNA into the inflorescences of barley plants
(*Hordeum vulgare*, cv. Erfa and Nebi). This technique was developed by De la Pena
et al. (1987) and was successfully applied to rye. 2,055 plants between the 3-leaf- and
6-leaf state were injected with a syringe; about 0.2 - 0.4 µg of CsCl-purified circular
plasmid DNA in a vol. of 5 - 10 µl were applied. Two constructs were used
containing the neomycin phosphotransferase II (*npt*) gene as selectable marker
thereby conferring resistance to kanamycin: pMLJ1glg, 12 kb, nos-promoter (Saalbach
et al., 1988) and pGSGLUC1, 14.2 kb, TR1´ promoter, containing additionally TR2´-
gus (kindly provided by Marc Van Montagu, Rijksuniversiteit and Plant Genetic
Systems) (*gus* = gene for ß-glucuronidase). The 28,300 seeds obtained were selected
on 150 mg/l kanamycin. 446 of them (= 1.6 %) grew in the presence of the antibiotic.

Untreated control plants grown on kanamycin showed to a small degree (0.6 %) also resistance to kanamycin.

All 446 selected seedlings were tested for NPT activity, and 45 of them showed reproducibly low activity levels corresponding to about 1 % of a stably transformed tobacco plant. These F1 plants were selfed and the progeny was shown to segregate on kanamycin, in some cases perfectly 3 : 1. However, in only very few F2 plants low NPT activity was detectable. It is under investigation whether the observed segregation is really linked to the introduced *npt* gene.

DNA isolated from F1 and F2 plants was further analysed. PCR primers specific for an internal 412 bp fragment of the *npt* coding sequence allowed the amplification of the target sequence in selected F1 and F2 plants also in those cases where no NPT activity was any longer detectable. When using the "left" PCR primer within the promoter region, pMLJ1glg-treated plants and pGSGLUC1-treated plants could be clearly differentiated. In pGSGLUC1-treated plants also a *gus*-specific 251 bp fragment was PCR-amplifyable.

Southern DNA hybridization of genomic DNA demonstrates that a hybridizing 1.7 kb npt-specific fragment is inherited from F1 to F2 plants, however different plants show surprisingly the same fragment size even when cutting with PstI inside the *npt*-coding sequence. Thus it appears that independent transformants show an identical hybridization pattern that is inherited from F1 to F2. We see no difference in hybridization patterns when comparing young seedlings with adult plants.

To exclude that we have transformed endophytes instead of the barley plant, vigorous attempts were made to isolate microbes from F1 and F2 leaves (with and without kanamycin in the medium): no microbes could be detected. Further, to exclude that we deal with extrachromosomal plasmid copies that have suddenly acquired the ability for self replication in plant cells, the isolated plant DNA was subjected to CsCl-gradient centrifugation: no plasmid bands could be observed. Furthermore, isolated plant DNAs were transformed into super-competent *E. coli* cells in order to detect free plasmid copies - with negative results.

A detailed comparative DNA analysis between different putative transformants on the F2 level is in progress in order to demonstrate unequivocally that DNA is integrated and to what extend vector sequences are included. The macroinjection approach was also used to introduce pGSGLUC1-DNA into wheat (cv. Chinese Spring). Both the *npt* and *gus* genes were inherited to the F2 where they were PCR-amplifyable and occurred non-rearranged on a 4 kb SalI-fragment (M. Dittmann and R.R. Mendel, unpublished).

DNA transfer using growing pollen tubes

We used the approach developed by Picard et al. (1988) to apply 3 µl CsCl-purified circular plasmid DNA (the same constructs as above) to the stigma of barley plants that were pollinated 5 - 20 min before DNA application. DNA was applied in a buffer mediating DNA stability for more than 1 hour: 1,058 plants treated; 11,200 seeds obtained and germinated on 150 mg/l kanamycin; 305 green plants selected

(= 2.7 %); 25 out of 101 plants already tested show low NPT-signals; F1 plants selfed to obtain the F2. Using PCR both the F1 and F2 show the *npt-* and promoter-specific sequences. Again, genomic DNA hybridization shows the presence of the 1.7 kb *npt-*specific fragment in the F1 and F2 plants tested.

Comparing these preliminary data, we see no principal difference between the frequencies and results of the pollen tube approach and the macroinjection procedure. In both cases it appears that a foreign DNA sequence is inheritable from F1 to F2 and in both cases the npt gene is only weakly expressed on the seedling level and loses its expression with plant growth to maturity and after inheritance to the next generation. This phenomenon of low expression is under further investigation. On the basis of NPT expression in the F1, a transformation frequency of about 10^{-3} to 10^{-4} could be deduced.

Transient gene expression in barley protoplasts

In order to show the principal expressibility in barley cells of the *npt* constructs used we determined at first their transient expression level. It turned out that TR1'-*npt*, CaMV35S-*npt* and nos-*npt* give high transient expression rates in barley mesophyll protoplasts. The same constructs were also tested in transient protoplast expression systems of sugarbeet cells (suspension culture) and *Nicotiana plumbaginifolia* (suspension culture).

This comparison shows that a promoter mediating strong expression in *N. plumbaginifolia* and sugarbeet cells gives also strong expression in barley cells, and vice versa. Hence, the low *npt* expression in the above described DNA-treated plants is not due to a general lack of expressibility of these constructs in barley.

Concerning transient *gus* gene expression in mesophyll as well as suspension culture protoplasts of barley, only low activities could be detected with TR1'-*gus* (corresponding to 1 - 10 % of that level which was observed in tobacco mesophyll protoplasts when using the same constructs). Also pRT103GUS (Töpfer et al., 1988, containing CaMV35S-*gus*) gave no higher activity.

DNA transfer into protoplasts

Suspension cultures were initiated from embryogenic calli derived from immature embryos, mesocotyle explants and inflorescences (cv. Borwina, Ilka, Golden Promise, Igri) and kept on MS-medium with 2 mg/l 2.4-D. Protoplasts isolated from these cultures could be readily regenrated to calli, in the case of Borwina with a plating efficiency of 25 %. These calli, however, did not regenerate.

About 106 protoplasts in a medium containing 75 mM calcium chloride and 60 mM sodium chloride were heat-treated (5 min 45 °C) followed by the addition of 20 µg circular DNA (pHP23 (kindly provided by J. Paszkowski, ETH Zürich, Switzerland), 4.4 kb, containing CaMV35S-*npt*) and PEG to a final conc. of 16 %. The protoplasts were imbedded into low-gelling agarose (1.5 %), selection on MS-medium containing 2 mg/l 2.4-D and 25 mg/l G418 started when the first microcolonies were visible (10 - 14 days after DNA-treatment). Colonies growing well after repeated subculturing on selective medium were tested for NPT activity. High levels of NPT activity were observed. PCR analysis of isolated DNA shows the presence of the expected 412 bp npt-specific fragment. In genomic Southern blots, *npt*-signals differing from the patterns found in the DNA-treated plants were detectable. Further DNA analysis comparing the latter with the former patterns is in progress. The transformation frequency was calculated to be approximately 10^{-4}.

DNA-transfer into suspension cells by particle gun

Suspension cells (cv. Borwina) were bombarded with DNA-coated tungsten microprojectiles (0.6 μm diameter) using a particle gun according to Sanford (Klein et al., 1987) (for review of this approach see Mendel, 1990). As plasmids, pHP23 (CaMV35S-*npt*), pGSGLUC1 (TR1´-*npt*, TR2´-*gus*) and pRT103GUS (CaMV35S-*gus*) were used and transient expression of these genes was reported previously by us (Mendel et al., 1989). Bombarded cells were subsequently selected on MS-medium containing 0.5 mg/l 2.4-D and 5 mg/l G418. The selected colonies showed only low NPT activities and no GUS activity (fluorimetric assay). DNA isolated from the selected cultures and subjected to PCR analysis showed reproducibly the *npt*-specific 412 bp fragment and the 251 bp *gus*-specific fragment, respectively. DNA hybridizations are in progress. Experiments using embryogenic targets have been started.

Conclusions

The results of all four approaches show that the *npt* gene under control of three different promoters is only weakly expressed in barley plants but shows higher expression in cell cultures. Hence a more suitable marker construct is needed for expression in barley plants. Nevertheless, the phenomenon of low expression is worth further investigation.

In a detailed further DNA analysis we will compare the hybridization patterns in blots of genomic DNAs isolated from the putative transformants that we obtained with the four approaches used. This will allow us to draw conclusions about the DNA integration when utilizing different techniques of direct DNA transfer.

The two cell genetic approaches offer the advantage of high numbers of target cells to be treated in a short time. However the calli obtained are not (yet) regenerable to plants. Thus the problem to obtain embryogenic calli derived from protoplasts has superior priority for the future. On the other hand, the two plant approaches give immediately fertile plants. The drawback with these approaches is, however, the high amount of manpower needed and the long time of the experiment. So the plant approaches represent principal solutions of the problem but are far away from any routine application.

References

De la Pena, A., Lörz, H. & Schell, J. (1987) *Nature* **325**, 274-276.

Klein, T.M., Bradley, A. & Fromm M.E. (1989) In: Genetic Engineering, Principles and Methods, Setlow, J.K.,(ed.), Vol. 11, Plenum Press, New York pp 13-31.

Klein, T.M., Wolf, E.D. & Sanford, J.C. (1987) *Nature* **327**, 70-73.

Mendel, R.R. (1990) *AgBiotech News and Information* **2**, 643-645.

Mendel, R.R., Clauss, E., Hellmund, R., Schulze, J., Steinbiß, H.H. & Tewes, A. (1990) In: Progress in plant cellular and molecular biology, Nijkamp, H.J.J. (ed.), Kluwer Academic Publishers, Dordrecht, pp 76-81

Mendel, R.R., Müller, B., Schulze, J., Kolesnikov, V. & Zelenin ,A. (1989) *Theor. Appl. Genet.* **78**, 31-34.

Picard, E., Jacquemin, J.M., Granier, F., Bobin, M. & Forgeois, P. (1988) Abstract Intl. Wheat Symposium, Cambridge UK.

Saalbach, G. et al. (1988) *Biochem. Physiol. Pflanzen* **183**, 211-218.

Töpfer, R., Pröls, M., Schell, J. & Steinbiß, H.H. (1988) *Plant Cell Rep.* **7**, 225-228.

Application of molecular analyses to questions relating to the genetics, ecology and evolution of actinorhizal symbioses

Beth C. Mullin, Susan M. Swensen, Paul Twigg and
Paula Goetting-Minesky

Department of Botany, Plant Physiology and Genetics Life Sciences Graduate Program and The Center for Legume Biology, The University of Tennessee, Knoxville, TN 37996-1100

Introduction

Actinorhizal symbiosis is the name given to the nitrogen fixing symbiotic relationship between actinomycetes of the genus *Frankia* and their woody host plants. Nitrogen fixed by frankiae in root nodules of host plants is transported throughout the infected plant allowing the plants to flourish on soils low in nitrogen. Host plants are distributed worldwide and actinorhizal symbioses are believed to play a major role in global nitrogen cycling as well as in the restoration and stabilization of disturbed lands (Schwintzer and Tjepkema, 1990; Baker and Mullin, In Press).

Ecological studies of *Frankia* have been hampered due to the difficulty in isolating *Frankia* from nodules and soil and in identifying specific bacterial strains. Studies on the evolution of the symbiosis have been hindered for want of firmly based hypotheses for the phylogeny of each partner in the symbiosis. Genetic analyses have been impeded by the lack of a system for transformation in *Frankia* and of host plant and symbiont mutants (Mullin and An, 1990). Molecular analyses are providing a basis for the identification and tracking of *Frankia* strains in the environment, for the construction of phylogenetic hypotheses, and for studies on gene expression.

Phylogenetic hypotheses related to host plants

One of the most striking features of actinorhizal symbioses is the taxonomic diversity of host plants. There are at least 160 known host plant species currently placed in six or seven angiosperm orders distributed among three or four subclasses. There is however no agreement among systematic botanists as to the proper placement of many of the actinorhizal genera. For example, the actinorhizal genus *Datisca* of the Datiscaceae has been allied with at least nine different families

(Begoniaceae, Capparaceae, Cistaceae, Cucurbitaceae, Flacourtiaceae, Haloragaceae, Loasaceae, Scyphostegiaceae and Violaceae) and there is no agreement as to its proper placement despite extensive morphological, anatomical and chemical analysis. There is even disagreement as to its placement at the subclass level.

The actinorhizal family Coriariaceae is composed of one genus, *Coriaria*, with five species. Coriariaceae has been placed in at least four distinct angiosperm orders (Sapindales, Celastrales, and Rosales of the Rosidae, and Ranunculales of the Magnolidae) and shares several morphological similarities with a fifth order, Trochodendrales of the Hamamelidae. There is also no agreement as to the ordinal or subclass level placement of this family despite extensive study.

The actinorhizal genera *Alnus, Allocasuarina, Casuarina,* and *Myrica* are considered by most systematists to be higher hamamelids. However, recently Doyle and Donoghue (1987) have suggested that the higher amentiferous hamamelids may actually be rosids based on the occurrence in this group of triporate pollen. This is a particularly interesting suggestion in light of the fact that the subclass Rosidae already contains the greatest number of actinorhizal genera.

These and other enigmatic systematic problems, related to the phylogeny of actinorhizal genera, are amenable to molecular analysis. It is possible that the analysis of nucleotide sequence data will help resolve conflicts surrounding the phylogenetic affinities of these plants.

Three separate but interactive genomes exist in plants; two organellar genomes, the chloroplast and mitochondria, and the nuclear genome. Mitochondrial genomes in plants are relatively large (200 kb to 2400 kb) and heterogeneous when compared to other eukaryotes (Sederoff, 1987). Mitochondrial DNA is often difficult to purify from nuclear and chloroplast DNA and for these reasons has not been well characterized (Palmer, 1985). Current macromolecular phylogenetic studies in plants are based largely on sequences from chloroplast or nuclear-encoded genes.

In the chloroplast genome the most widely sequenced gene is rbcL which encodes the large subunit of ribulose-1,5-bisphosphate carboxylase. This slowly evolving gene has become the preferred gene for the investigation of higher-level phylogenetic relationships such as the ones discussed here. Comparative sequencing of rbcL has been recently employed to help resolve many issues in angiosperm phylogeny, such as the systematics of the Caryophyllidae (Rettig et al., 1990), interfamilial relationships of the Asteraceae (Michaels and Palmer, 1990) and phylogenetic relationships in Saxifragaceae (Soltis et al., 1990). At the time of this symposium the rbcL gene has been sequenced from at least 200 species of plants (Palmer, personal communication).

In addition to rbcL, nuclear genes encoding ribosomal RNA (rRNA) have been utilized for plant phylogenetic analysis. Different rates of evolution occur in different regions of the ribosomal gene which allow the gene to be used for either low level phylogenetic comparisons (subgeneric) or higher level inter or intrafamilial comparisons (Hamby and Zimmer, 1991). Several studies of early angiosperm evolution have used rRNA sequences to construct phylogenetic hypotheses (Hamby, 1990; Wolfe et al., 1989; Nairn and Ferl, 1988) and rRNA sequences from at least 60 plant species are available for comparison.

We have embarked on a study directed towards resolving conflicts surrounding the phylogenetic placement of several actinorhizal genera. Both rbcL and rRNA sequences are being generated from representative actinorhizal species, species from postulated relative groups and appropriate outgroups. Several different methods are being used to construct phylogenies based on sequence data because no single method is recognized as the optimal method and the resulting phylogenies will be evaluated by extensive statistical testing.

Hypotheses on the evolution of actinorhizal symbioses have previously been made (Mullin et al., 1990), but it is clear that further consideration of such hypotheses must await the construction of firmly based host plant phylogenies. We expect that molecular analyses will provide the data necessary for phylogenetic reconstruction.

Genetics of the host plants

Although there are variations in the mechanism of infection by *Frankia* and in nodule morphology and physiology among host plants, all actinorhizal nodules have an internal anatomy similar to that of a lateral root. Nodule lobe primordia are initiated from the root pericycle and grow outward through the root cortex. Nodule development is a plant developmental process induced by invading *Frankia* and would be expected to exhibit differential gene expression in a manner that has been demonstrated in many other plant organs.

One approach to studying differential gene expression in actinorhizal nodules is to measure the level of specific mRNA transcripts in nodules at different developmental stages. *In situ* hybridization to nodule sections using mRNA directed probes would allow a cell to cell comparison to be made as well. In order to prepare such probes we have constructed both root and nodule cDNA libraries and from these we have made a subtraction library that should be enriched in cDNAs for nodule specific or nodule enhanced mRNA transcripts.

In addition to studying differential gene expression in host plants in response to *Frankia*, we are interested in the status of these genes in non-host plants and the extent to which non-host plants are able to recognize and respond to *Frankia*. *Betula*, a non-actinorhizal genus placed in the same family with actinorhizal *Alnus* species, has been shown by restriction fragment analysis of rDNA to be very closely related to *Alnus*. In fact, cluster analysis using the unweighted pair group method using arithmetic averages (UPGMA, Sneath and Sokal, 1973), clusters *Betula* with two *Alnus* species in a group distinct from the remaining five *Alnus* species tested (Bousquet et al., 1989). Other studies have shown that in some locations soil under non-host *Betula* induces nodulation in actinorhizal *Alnus incana* to a greater extent than soil under nearby *Alnus* (Smolander, 1990). *Betula* has been proposed as a target for the genetic engineering of new nitrogen fixing associations through gene transfer (Bousquet et al., 1989) but to date only marker genes have been inserted into the *Betula* genome (Mackay et al., 1988; Seguin and Lalonde, 1990). Identification and study of the expression of genes found to be important in the *Alnus* symbiosis will target sequences to be used in transformation experiments aimed at establishing new symbiotic relationships.

Phylogeny of *Frankia*

The earliest attempts to determine the phylogenetic position of *Frankia* among the actinomycetes based on morphological and chemical analyses supported its alignment with the genus *Dermatophilus*, an obligate animal pathogen not able to survive in soil (Lechevalier and Lechevalier, 1979). The most rudimentary molecular analysis, determination of the G+C content of *Frankia* DNA, showed that this relationship was not representative of the true phylogeny of the organisms. The G+C content of *Frankia* strains ranged between 10-15% higher than the G+C content of *Dermatophilus* (An et al., 1983). Although DNA solution hybridization of *Frankia* DNA to that of a wide range of actinomycetes did not reveal any close relatives (An et al., 1987), the more sensitive analyses of rRNA by oligonucleotide cataloguing and sequencing identified *Geodermatophilus* as a distant but nearest relative (Fox and Stackabrandt, 1987; Hahn, et al., 1989b). *Geodermatophilus*, a soil actinomycete was intuitively a much better candidate for a nearest relative than the obligate animal pathogen *Dermatophilus*.

With the genus *Frankia* firmly placed among the actinomycetes, the question arises as to the origin of its nitrogen fixing and symbiotic genes. Nearest relatives are not known to form any kind of association with plants nor are they known to fix nitrogen. In fact few, if any, other actinomycetes have been observed to fix nitrogen. DNA sequence analysis of the structural genes for the enzyme nitrogenase has provided some insight into the origin of these genes in *Frankia*. We have sequenced *nif*D, the structural gene for the α subunit of the MoFe protein of the nitrogenase complex, and found it to have relatively higher sequence similarity to *nif*D from *Anabaena* than to other nitrogen-fixing bacteria (Twigg et al., 1990). *nif*H from two *Frankia* strains also has relatively higher sequence similarity to *nif*H from *Anabaena* sp. (Normand and Bousquet, 1989). In both cases the corresponding sequences from *Clostridium* share much less similarity with *Frankia* despite the fact that these two genera are more closely related based on rRNA sequence data. Further analysis of the organization and sequence of genes which function in nitrogen fixation will strengthen or repudiate the speculation of Simonet et al. (1990) that *Frankia* may have obtained its genes for nitrogen fixation via horizontal gene transfer.

Ecological Studies

Despite the general presence of *Frankia* in the soil little is known about the ability of *Frankia* to grow and compete in the soil. There is little information about strain competition for nodulation of host plants. Genetic analysis of *Frankia* isolates indicates that there is genetic diversity within single localities, and physiological analysis has shown that isolates differ significantly from one another in symbiotic properties. It has not been possible to routinely isolate frankiae from the soil, and nodule isolates, although more readily obtained, cannot be made with consistency. Furthermore distinguishing among isolates once they are available is also problematic because there are few phenotypic characters on which to base identification.

Once again molecular analyses have provided both the means to identify and to track *Frankia* strains. Although the systematics of frankiae is in its infancy, the most promising way to group strains to date has been based on DNA solution hybridization (An et al., 1985a; Fernandez et al., 1989). No species names are yet in wide use but potential species groups have been identified by this method. Distinction among pure culture strains has been possible by observation of total restriction fragment patterns (An et al., 1985b; Dobritza, 1985; Ide, 1986; Bloom et al., 1989) but this method is of little value for identifying strains within nodules. RFLP hybridization patterns used by Nittayajarn et al. (1990) to screen for host specificity of pure culture strains are of potential value in determining nodule occupancy. DNA probes specific to the *Frankia* DNA are being used to probe Southern blots of total nodule DNA. In collaboration with Baker and Nittayajarn (Yale School of Forestry) we have obtained hybridization bands which can be used to identify *Frankia* strains.

Strain-specific DNA probes would provide the most practical way to track specific *Frankia* strains. These probes could be hybridized to total nodule DNA obtained simply by crushing a nodule. The result would be a positive or negative signal. The first application of a *Frankia* strain-specific probe involved the use of a plasmid probe to map the presence or absence of the plasmid among actinorhizal nodules in a limited geographical area (Simonet et al., 1988). In this study, direct hybridization to nodule extracts was not specific and the extract had to be electrophoresed before specificity was observed. The broader use of plasmids probes to track *Frankia* strains should be undertaken with caution because of the lack of correlation that can exist between the presence of a plasmid and the genotype of the frankiae (Bloom et al., 1989). Ribosomal RNA is an abundant molecule present in cells at a much higher concentration than are its genes. Hahn et al. (1989a; 1990) have taken advantage of this fact, coupled with the intrinsic variation that exists within the RNA molecule, to create strain-specific oligonucleotide probes. These have a tremendous potential to facilitate the tracking of specific strains.

For studies that do not involve pure culture strains or if no strain-specific probe is available, direct sequencing of PCR-amplified variable regions of *Frankia* genes would provide a way to assess diversity within and nodulation frequencies of a population. Using conserved sequences in the 3' region of *nif*H and the 5' region of *nif*D as primers we have amplified the intergenic region between *nif*H and *nif*D which we had previously found by sequence analysis to be variable among *Frankia* strains.

Techniques such as these will be useful in determining the structure of *Frankia* populations in the soil and the effects that host and non-host plants have on population dynamics and strain competition in natural environments. None of these studies would be possible without the application of molecular analyses.

References

An, C.S., Wills, J.W., Riggsby, W.S. & Mullin, B.C. (1983) *Canadian J. Botany* 61, 2859-2862.

An, C.S., Riggsby, W.S. & Mullin, B.C. (1985a) *Intl. J. Syst. Bacteriol.* 35, 140-146.

An, C.S., Riggsby, W.S. & Mullin, B.C. (1985b) *Plant and Soil* 87, 43-48.

An, C.S., Riggsby, W.S. & Mullin, B.C. (1987) *The Actinomycetes* 20, 50-59.

Baker, D.D. & Mullin, B.C. (In Press) In: Biological Nitrogen Fixation, Stacey, G., Evans, H. and Burris, R. (eds), Chapman and Hall, New York, London.

Bloom, R.A., Mullin, B.C. & Tate, R.L., III (1989) *Appl. Environ. Microbiol.* 55, 2155-2160.

Bousquet, J., Girouard, E., Strobeck, C., Dancik, B.P. & Lalonde, M. (1989) *Plant and Soil* 118, 231-240.

Dobritsa, S.V. (1985) *FEMS Microbiol. Lett.* 29, 123-128.

Doyle, J.A. & Donoghue, M.J. (1987) *Review of Paleobotany and Palynology*, 50, 63-95.

Fernendez, M.P., Meugnier, H., Grimont, P.A.D. & Bardin, R. (1898) *Int. J. Syst. Bacteriol.* 39, 424-429.

Fox, G.E. & Stackebrandt, E. (1987) *Methods in Microbiol.* 19, 405-458.

Hahn, D., Dorsch, M., Stackebrandt, E. (1989a) *Plant and Soil* 118, 211-219.

Hahn, D., Lechevalier, M.P., Fischer, A., & Stackebrandt, E. (1989b) *System. App. Microbiol.* 11, 236-242.

Hahn, D., Starrenburg, M.J.C., Akkermans, A.D.L. (1990?) *Appl. Environ. Microbiol.*

Hamby, R.K. (1990) Ph.D. Dissertation. Lousiana State University.

Hamby, R.K. & Zimmer, E.A. (1991) In: Plant Molecular Systematics, Soltis, D. Soltis, P. & Doyle, J. (eds). In Press.

Ide, P.I. (1986) M.S. Thesis, The University of Tennessee, Knoxville.

Lechevalier, M.P. & Lechevalier, H.A. (1979) In: Symbiotic Nitrogen Fixation in the Management of Temperate Forests, Gordon, J.C., Wheeler, C.T. & Perry, D.A. (eds), Forestry Sciences Laboratory, Oregon State University, Corvallis, OR.

Mackay, J., Seguin, A. & Lalonde, M. (1988) *Plant Cell Reports* 7, 229-232.

Michaels, H.J. & Palmer J.D. (1990) *Am. J. Bot.* 77(6), 177.

Mullin, B.C. & An, C.S. (1990) In: The Biology of Frankia and Actinorhizal Plants, Schwintzer, C.R. and Tjepkema, J.D. (eds), Academic Press, San Diego, New York, pp 195-214.

Mullin, B.C., Swensen, S.M. & Goetting-Minesky, M.P. (1990) In: Nitrogen Fixation: Achievements and Objectives, Gresshoff, P.M., Roth, L.E., Stacey, G. & Newton, W.E. (eds), Chapman & Hall, New York, London, pp 781-787.

Nairn, C.J. & Ferl, R.J. (1988) *J. Mol. Evol.* 27, 133-141.

Nittayjarn, A., Mullin, B.C. & Baker, D.D. (1990) *Appl. Environ. Microbiol.* 56, 1172-1174.

Normand, P. and Bousquet, J. (1989) *J. Mol. Evol.* 29, 436-447.

Palmer, J.D. (1985) In: Molecular Evolutionary Genetics, MacIntyre (ed), Plenum Press, New York.

Rettig, J.H., Wilson, H.D. & Manhart (1990) *Am. J. Bot.* 77(6), 177.

Schwintzer, C.R. and Tjepkema, J.D. (1990) The Biology of Frankia and Actinorhizal Plants, Academic Press, San Diego, New York.

Sederoff, P.R. (1987) *American Naturalist* 130, S30-S45.

Seguin, A. & Lalonde, M. (1990) In: The Biology of Frankia and Actinorhizal Plants, Schwintzer, C.R. & Tjepkema, J.D. (eds), Academic Press, New York, pp 215-238.

Simonet, P., Normand, P., Hirsch, A.M. & Akkermans, A.D.L. (1990) In: The Molecular Biology of Symbiotic Nitrogen Fixation, Gresshoff (ed), CRC Press, Boco Raton, FL, pp 77-109.

Simonet, P., Thi Le, N., Teissier du Cros, E. & Bardin, R. (1988) *Appl. Environ. Microbiol.* 54, 2500-2503.

Smolander, A. (1990) *Plant and Soil* 121, 1-10.

Sneath, P.H.A. & Sokal, R.R. (1973) Numerical Taxonomy, W. H. Freeman, San Francisco.

Soltis, D.E., Soltis, P.S., Clegg, M.T. & Durbin, M. (1990) *Am. J. Bot.* 77(6), 118.

Twigg, P., An, C.S. & Mullin, B.C. (1990) In: Nitrogen Fixation: Achievements and Objectives, Gresshoff, P.M., Roth, L.E., Stacey, G. & Newton, W.E., (eds), Chapman & Hall, Publ. New York, London.

Wolfe, K.H., Gouy, M., Yang, Y.-N., Sharp, P.M. & Li, W.-H. (1989) *PNAS USA* 86, 6201-6205.

Intracellular receptor proteins for calcium signals in plants

Daniel M. Roberts, C. David Weaver, and
Suk-Heung Oh

*Department of Biochemistry and Center for Legume Research, The University
of Tennessee, Knoxville, TN 37996-0840*

Introduction

From the intensive research on calcium homeostasis and stimulus response coupling it is clear that calcium signal transduction systems are among the most universal in eukaryotic cells. In higher plants, calcium signals have been implicated in the control of a diverse number of cellular responses that are linked to plant growth and development (reviewed by Hepler and Wayne, 1985; Poovaiah and Reddy, 1987). Studies of animal model systems suggest that the nature and control of the calcium signal is complex (Berridge and Irvine, 1989; Rasmussen, 1990). Although much less is known about the calcium signal in plant cells, recent work has suggested that certain components of the calcium signalling pathway, such as the phosphoinositide cycle and inositol-tris phosphate responsive channels, are present in plants (see Einspahr and Thompson, 1990 for a review).

In order to understand the mechanism of how these calcium signals facilitate cellular responses in plants it is necessary to consider the intracellular calcium receptors, the calcium-modulated proteins. Over 100 different calcium-modulated proteins have been sequenced, again supporting the diversity and complexity of the calcium signalling apparatus (Moncrief et al., 1990). Two major receptors of calcium regulatory signals have been characterized in plants: calmodulin and a new, novel calcium-dependent protein kinase. As discussed below, both of these proteins are widely distributed in plants and are likely to be play a multifunctional role in calcium signalling.

Calmodulin

Calmodulin is probably the most well studied intracellular calcium regulatory protein (for recent reviews see Cohen and Klee, 1988; Roberts et al., 1986). Calmodulin is a single polypeptide chain that binds four calcium ions reversibly ($Kd = 10^{-6}$ M), and activates a number of enzymes in a calcium-dependent manner. The essential role that calmodulin plays as an intracellular receptor of calcium signals becomes clear when one considers its ubiquitous distribution and the fact that the deletion of the calmodulin allele is a lethal event (Davis et al., 1986; Takeda and Yamamoto, 1987).

Although it has been 14 years since the first description of calmodulin in higher plant tissues, a clear mechanistic picture of the calmodulin regulatory system in plants has yet to emerge. Selected cellular functions that are proposed to involve regulation by calmodulin are listed in Table 1. Observations that calmodulin levels can be significantly modulated in response to growth stimuli (Braam and Davis, 1990) and that this protein is specifically associated with intracellular structures such as microtubules during cell division (Wick, 1989), argue for a key role of calmodulin in growth regulation of plants. However, defining this role is the difficult challenge that awaits calmodulin researchers. Especially important will be studies of the targets of calmodulin action: calmodulin-dependent enzymes and calmodulin-binding proteins.

Table 1 Selected plant processes proposed to involve calmodulin regulation.

NAD kinase/nicotinamide co-enzyme homeostasis	Muto & Miyachi, 1977; Anderson & Cormier, 1978.
calcium ATPases/calcium homeostasis	Dieter & Marme, 1981.
kinetochore microtubule function	Wick, 1989.
nuclear NTPases	Chen et al., 1987.
thigmomorphogenesis	Braam & Davis, 1990.

Very few calmodulin target proteins have been identified in plant tissues. Probably the most well characterized of these is the enzyme NAD kinase. Activation of this enzyme by calmodulin is thought to be involved in an oxidative or metabolic "burst" of activity in stimulated cells. For example, in animal model systems such as sea urchin, it is proposed that calmodulin-dependent NAD kinase is activated upon fertilization, providing a burst of NADPH for the metabolic activation of the fertilized egg (Epel et al., 1981). It remains unknown, however, what specific metabolic processes are affected by calmodulin-dependent NAD kinase in higher plant tissues.

As one approach to this system, we have been investigating the distribution of the enzyme in higher plants. In Fig. 1 the distribution of NAD kinase in nitrogen-fixing soybean plants is shown. Surprisingly, very low levels of activity were detected in the leaves, and this activity tended to be calcium independent. The highest NAD kinase specific activities were found in the root and nodule tissues.

Nearly all of the activity detected in the nodule was calcium dependent (Fig. 2). This finding is of interest because other investigators have claimed that two separate

NAD kinase activities exist in plants: a calcium and calmodulin-dependent activity and a calcium-independent activity (Simon et al., 1984; Muto and Miyachi, 1986).

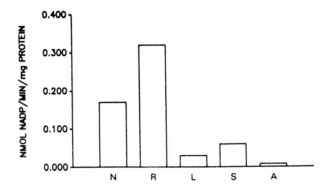

Figure 1 Distribution of NAD kinase enzyme activity in nitrogen-fixing soybean plants. 28 day old soybean plants infected with Bradyrhizobium japonicum *strain 61A101C were dissected into nodules (N), roots (R), leaves (L), stems (S), and apices (A), and were extracted and tested for NAD kinase activity (nmole NADP produced per min per mg soluble protein).*

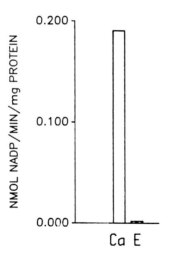

Figure 2 Calcium dependence of nodule NAD kinase. Extracts of nodules from 28-day soybean plants were assayed in the presence of 1 mM CaCl$_2$ (Ca) or 1 mM EGTA (E).

The finding that the major NAD kinase activity found in the nodules is calcium dependent argues that nicotinamide co-enzyme fluxes in this organ are principally under the control of calcium and calmodulin. Considering the differential utilization of NAD and NADP in energy generating and biosynthetic pathways, one might envision that the control of this enzyme could influence the balance of nicotinamide co-enzymes which could lead to changes in metabolic fluxes within the nodule.

One additional consideration in the regulation of NAD kinase is the influence of posttranslational modification of calmodulin on NAD kinase activation. We have previously shown that posttranslational methylation of calmodulin results in a lower overall activation of NAD kinase, but does not influence other calmodulin-dependent enzymes (Roberts et al., 1984; 1985; 1990). An example of this is illustrated in Table 2 with NAD kinase purified from soybean nodules. This observation combined with the recent finding that the levels of posttranslational calmodulin methylation can vary depending on the developmental state of the plant tissue (Oh and Roberts, 1990) has led to the proposal that NAD kinase activation could be selectively attenuated by posttranslational methylation of calmodulin at lysine 115. Recently, we have found that transgenic plants that express calmodulin mutants that are incapable of methylation and that hyperactivate NAD kinase show specific growth deficiencies (Oh et al., 1991). By using this combined genetic and biochemical approach we hope to gain further insight into NAD kinase regulation in the plant, and the biological importance of calmodulin methylation.

Table 2 Calmodulin dependence of soybean nodule NAD kinase[a]

Treatment	NAD kinase activity
1 mM EGTA	0
1 mM CaCl$_2$	5
1 mM CaCl$_2$/spinach CaM[b]	43 (1.6)[c]
1 mM CaCl$_2$/VU-1 CaM 84	(8)

a. Soybean nodule NAD kinase was purified through the DEAE cellulose step as described by Roberts et al., 1985. NAD kinase activity is measured as nmole NADP produced per min.

b. 100 nM of spinach calmodulin (Lukas et al., 1984) or VU-1 calmodulin (Roberts et al., 1985) was used in each assay. Spinach calmodulin is trimethylated at lysine 115. VU-1 calmodulin is unmethylated.

c. Numbers in parentheses represent the standard error of the mean.

Calcium-dependent, calmodulin-independent protein kinases

Among the multitude of cellular targets of calcium in eukaryotes are protein kinase enzymes that modulate a variety of cellular activities through the reversible phosphorylation of substrate proteins (see Edelman et al., 1987; Hanks et al., 1988 for reviews). In animal tissues the effects of calcium on protein kinase activities is principally mediated through calmodulin-dependent protein kinases or lipid-dependent enzymes such as protein kinase C (Edelman et al., 1987). Although several investigators have attempted the identification of similar enzymes in plant tissues, there has yet to be an unequivocal demonstration of calmodulin-dependent protein kinases or protein kinase C in higher plants. However, the presence of a novel type of calcium-dependent protein kinase recently has been reported in several higher plant (Harmon et al., 1987; Battey and Venis, 1988; Putnam-Evans et al., 1990; Weaver et al., 1991), algal (Roberts, 1989; Guo and Roux, 1990), and protist (Gundersen and Nelson, 1987) species. Our recent focus with this enzyme has been on the investigation of its potential role in the phosphorylation of the nodule-specific, symbiosome membrane protein nodulin 26 (Weaver et al., 1991).

<u>NODULIN 26</u>

```
1                                                             33
M A D Y S A G T E S Q E V V V N V T K N T S*E T I Q R S D S*L V S
34                                                            66
L P F L Q K L V A E A V G T Y F L I F A G C A S L V V N E N Y Y N
67                                                            99
M I T F P G I A I V W G L V L T V L V Y T V G H I S G G H F N P A
100                                                          132
V T I A F A S T R R F P L I Q V P A Y V V A Q L L G S I I A S G T
133                                                          165
L R L L F M G N H D Q F S G T V P N G T N L Q A F V F E F I M T F
166                                                          198
F L M F V I C G V A T D N R A V G E L A G I A I G S T L L L N V I
199                                                          231
I G G P V T G A S M N P A R S L G P A F V H G E Y E G I W I Y L L
232                                                          264
A P V V G A I A G A W V Y N I V R Y T D K P L S*E I T K S A S*E L
265              271
K G R A A S*K
```

<u>CK-15</u>

```
C T K S A S F L K G R A A S K
```

Figure 3 Amino acid sequence of nodulin 26 and CK-15. The deduced amino acid sequence of nodulin 26 (Fortin et al., 1987; Sandal and Marcker, 1988) and the amino acid sequence of CK-15 are shown. The region of homology between the COOH terminus of nodulin 26 and CK-15 is underlined. Asterisks show potential sites (serine residues 22, 30, 255, 262, and 270) of phosphorylation based on previous studies with Ca^{2+}-dependent protein kinases (Kemp et al., 1983; Stull et al., 1986; Roberts, 1989).

The amino sequence deduced from the nodulin 26 cDNA (Fortin et al., 1987; Sandal and Marcker, 1988) is shown in Figure 3. This protein has been found to have limited (30 to 40 %) sequence homology to a number of plant, bacterial and animal membrane proteins that are thought to have channel activity (Pao et al., 1990). Thus, it is proposed that nodulin 26 may serve a function in metabolite transport on the symbiosome membrane (Sandal and Marcker, 1988, 1990; Verma, 1990).

An examination of the nodulin 26 sequence (Fig. 3) revealed several potential phosphorylation motifs (R/K-X-X-S/T, with X representing any amino acid) for calcium-dependent protein kinases within the hydrophilic amino-terminal and carboxyl-terminal domains. Interestingly, another member of the nodulin 26 family, the major intrinsic protein of bovine lens, has been shown to undergo phosphorylation at homologous residues within the carboxyl terminal domain (Lampe and Johnson, 1990). We pursued the potential phosphorylation of nodulin

26 by synthesizing the CK-15 peptide shown in Fig. 3. This peptide was selected since it contains two potential phosphorylation sites, and also because it is hydrophilic and unique in sequence to nodulin 26, and thus can be used for the production of site-directed, nodulin 26 specific antibodies.

Table 3 *Detection of CK-15 protein kinase activity in 28-day old nitrogen-fixing root nodules.*

Substrate	kinase activity [a]	
	Ca	EGTA
none	0.073	0.051
130 μM CK-15	0.894	0.164

a. Kinase activity was measured in nmole phosphate incorporated per minute per mg protein. Standard phosphocellulose filter assays were done (Roberts et al., 1984). Ca, 1 mM $CaCl_2$; EGTA, 1 mM EGTA.

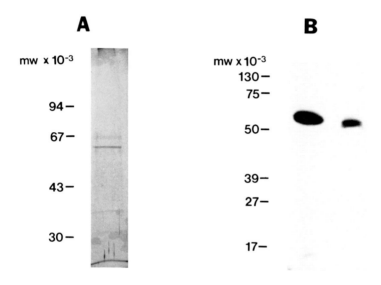

Figure 4 *SDS-polyacrylamide gel electrophoretic analysis of CK-15 protein kinase from soybean nodules. The enzyme was purified as previously described (Weaver et al., 1991). A. silver-stained gel of purified CK-15 protein kinase from nodules; B. autoradiogram of a protein kinase activity stain with histone H1 as substrate (Geahlen et al., 1986). Lane 1 of panel B contains the calcium-dependent protein kinase from the green alga* Mougeotia *(Roberts, 1989), and lane 2 contains the calcium-dependent CK-15 protein kinase from soybean nodules.*

Crude nodule extracts show a calcium-dependent protein kinase activity that phosphorylates the CK-15 peptide (Table 3). This peptide-based assay was used to purify the CK-15 protein kinase. SDS-polyacrylamide gel electrophoresis of the final product showed a major polypetide band with an apparent molecular weight of

60,000 (Fig. 4A). A protein kinase activity stain shows that this same band possesses protein kinase activity (Fig. 4B).

The enzyme shows complete calcium-dependent activity (Fig. 5) with a $K_{0.5}$ in the physiological range (i.e., μM). The presence of calmodulin or protein kinase C effectors (diacylglycerol and phosphatidyl serine) has no significant influence on the activity of the enzyme. Thus, it has been proposed that calcium a direct effector of the enzyme, and that it activates the enzyme by directly interacting with a calcium-binding regulatory domain on the catalytic protein kinase subunit (Gunderson and Nelson, 1987; Harmon et al., 1987; Harper et al., 1991). This mechanism of activation is novel among all calcium-dependent protein kinases described thus far.

By using CK-15, and related synthetic peptide substrates, we have detected this calcium-dependent protein kinase activity in all plant and algal tissues we have tested (Roberts, 1989; Weaver et al., 1991). Analysis of purified kinases from some of these biological sources show that they have properties similar to those of the soybean nodule CK-15 kinase (Roberts, 1989; Fig. 4B). Thus, the protein kinase activity that phosphorylates the CK-15 sequence is widely distributed in the plant kingdom and may play a pleiotropic and multifunctional role in calcium-dependent regulation of biological processes in plants.

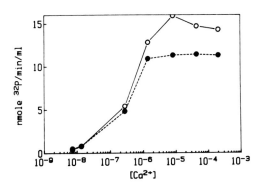

Figure 5 Calcium dependence of purified CK-15 protein kinase from root nodules. The enzyme was assayed under standard protein kinase assay conditions (Weaver et al., 1991) in the presence of various concentrations of calcium ions and the presence (open circles) or absence (closed circles) of 24 µM phosphatidyl serine and 0.64 µM 1,2-dioleoylglycerol.

Summary and Prospects

Calmodulin and calcium-dependent protein kinases both represent integral calcium-regulatory targets that are widespread in plant tissues. Considering the diversity of roles that calcium signals play in plant cellular processes, the ubiquitous nature of these calcium receptor proteins is not surprising. However, the specific mechanisms through which these calcium-target proteins mediate calcium signal transduction remains elusive. In this regard it is important to consider the processes that occur in the next step of the signal transduction pathway following calcium binding. For example, in the case of calmodulin, its role will be defined by the types and quanitities of calmodulin-dependent enzymes and calmodulin-binding proteins found within a given cell. Clearly, the finding that NAD kinase is a principal target of calmodulin in higher plants argues for a role in the control of metabolism by regulating nicotinamide co-enzyme fluxes. However, specific details of this role are lacking. Additionally, considering that multiple calmodulin-dependent enzymes have been defined in animal systems, there are potentially several other calmodulin-target proteins in plants that have not yet been discovered.

In the case of the calcium-dependent protein kinase, its role within a particular cell will be defined by the types and quantities of the endogenous substrates found within that cell. Thus, it is important to elucidate the *in vivo* sustrates of this enzyme and to assess the influence of phosphorylation on their activity. Little progress has been made in this area. However, it is clear from the work summarized above that nodulin 26 is a likely target of this enzyme, and similar approaches may identify other potential substrates in other tissues. With respect to nodulin 26, further work is required to assess the role of phosphorylation on its function, and to determine the nature of the signals that control its phosphorylation *in vivo*.

As one final note, it is interesting to consider that calmodulin and the calcium-dependent protein kinase represent two distinct and yet structurally and functionally similar calcium regulatory proteins. Studies with animal systems have shown that the calcium regulatory machinery is complex and consists of a myriad of calcium binding proteins that co-exist and operate within the same cell in a concerted fashion. Thus, with respect to the plant cell, calmodulin and the calcium-dependent protein kinase may represent only two of many calcium receptors that remain to be discovered in plant tissues.

Acknowledgement

Supported by NSF grant DMB 87-16224 and USDA grant 88-37261-3521. We wish to acknowledge the technical assistance of Burnette Crombie in this work.

References

Anderson, J.M. & Cormier, M.J. (1978) Biochem. Biophys. Res. Commun. **84**, 595-602.

Battey, N.H. & Venis, M.A. (1988) Anal. Biochem. **170**, 116-122.

Berridge, M.J. & Irvine, R.F. (1989) Nature **341**, 197-204.

Braam, J. & Davis, R.D. (1990) Cell **60**, 357-364.

Chen, Y.-R., Datta, N. & Roux, S.J. (1987) J. Biol. Chem. **262**, 10689-10694.

Cohen, P. & Klee, C.B. (1988) Calmodulin. Molecular Aspects of Cellular Regulation, Vol. 5, Elsevier Biomedical Press, Amsterdam.

Davis, T.N., Urdea, M.S., Masiarz, F.R. & Thorner, J. (1986) Cell **47**, 423-431.

Dieter, P. & Marme, D. (1981) FEBS Lett. **125**, 245-248.

Edelman, A.M., Blumenthal, D.K. & Krebs, E.G. (1987) Ann. Rev. Biochem. **56**, 567-614.

Einspahr, K.J. & Thompson, G.A. (1990) Plant Physiol. **93**, 361-366.

Epel, D., Patton, C., Wallace, R.W. & Cheung, W.Y. (1981) Cell **23**, 543-549.

Fortin, M.G., Morrison, N.A. & Verma, D.P.S. (1987) Nucl. Acids Res. **15**, 813-824.

Geahlen, R.L., Anostario, M., Low, P.S. & Harrison, M.L. (1986) Anal. Biochem. **153**, 151-158.

Gundersen, R.E. & Nelson, D.L. (1987) J. Biol. Chem. **262**, 4602-4609.

Guo, Y.-L. & Roux, S.J. (1990) Plant Physiol. **94**, 143-150.

Hanks, S.K., Quinn, A.M. & Hunter, T. (1988) Science **241**, 42-52.

Harmon, A.C., Putnam-Evans, C. & Cormier, M.J. (1987) Plant Physiol. **83**, 830-837.

Harper, J.F., Sussman, M.R., Schaller, G.E., Putnam-Evans, C., Charbonneau, H. & Harmon, A.C. (1991) Science **252**, 951-954.

Hepler, P.K. & Wayne, R.O. (1985) Ann. Rev. Plant Physiol. **36**, 397-439.

Kemp, B.E., Pearson, R.B. & House, C. (1983) Proc. Natl. Acad. Sci. U.S.A. **80**, 7471-7475.

Lampe, P.D. & Johnson, R.G. (1990) Eur. J. Biochem. **194**, 541- 547.

Lukas T.J., Iverson, D.B., Schleicher, M. & Watterson, D.M. (1984) Plant Physiol. **75**, 788-795.

Moncrief, N.D., Kretsinger, R.H. & Goodman, M. (1990) J. Mol. Evol. **30**, 522-562.

Muto, S. & Miyachi, S. (1977) Plant Physiol. **59**, 55-60.

Muto, S. & Miyachi, S. (1986) In: Molecular and Cellular Aspects of Calcium in Plant Growth and Development, Trewavas, A. (ed.), Plenum Press, New York, pp. 107-114.

Oh, S.-H. & Roberts, D.M. (1990) Plant Physiol. **93**, 880-887.

Oh, S.-H., Besl, L., Stacey, G. & Roberts, D.M. (1991) FASEB J. **5**, A449.

Pao, G.M., Wu, L.-F., Johnson, K.D., Hofte, H., Chrispeels, M.J., Sweet, G., Sandal, N.N. & Saier, M.H. (1991) Mol. Microbiol. **5**, 33-37.

Poovaiah, B.W. & Reddy, A.S.N. (1987), CRC Crit. Rev. Plant Sci. **6**, 47-103.

Putnam-Evans, C.L., Harmon, A.C. & Cormier, M.J. (1990) Biochemistry **29**, 2488-2495.

Rasmussen, H. (1990) Biol. Chem. Hoppe-Seyler **371**, 191-206.

Roberts, D.M., Burgess, W.H. & Watterson, D.M. (1984) Plant Physiol. **75**, 796-798.

Roberts, D.M., Crea, R., Malecha, M., Alvarado-Urbina, G., Chiarello, R.H. & Watterson, D.M. (1985) Biochemistry **24**, 5090-5098.

Roberts, D.M., Lukas, T.J. & Watterson, D.M. (1986), CRC Crit. Rev. Plant Sciences **4**, 311-339.

Roberts, D.M. (1989) Plant Physiol. **91**, 1613-1619.

Roberts, D.M., Oh, S.-H., Besl, L., Weaver, C.D. & Stacey, G. (1990) In: Current Topics in Plant Biochemistry and Physiology, Randall, D.D. & Blevins, D.G. (eds.), University of Missouri, Columbia, Vol. 9, pp. 67-84.

Sandal, N.N. & Marcker, K.A. (1988) Nucl. Acids Res. **16**, 9347.

Sandal, N.N. & Marcker, K.A. (1990) In: Nitrogen Fixation: Achievements and Objectives, Gresshoff, P.M., Roth, L.E., Stacey, G. & Newton, W.E. (eds.), Chapman & Hall, New York, pp. 687-692.

Simon, P., Bonzon, M., Greppin, H. & Marme. D. (1984) FEBS Lett. **167**, 332-337.

Stull, J.T., Nunnally, M.H. & Michnoff, C.H. (1986) The Enzymes **17**, 114-166.

Takeda, T. & Yamamoto, M. (1987) Proc. Natl. Acad. Sci. U.S.A. **84**, 3580-3584.

Verma, D.P.S. (1990) In: Nitrogen Fixation: Achievements and Objectives, Gresshoff, P.M., Roth, L.E., Stacey, G., & Newton, W.E. (eds.), Chapman & Hall, New York, pp. 235-237.

Weaver, C.D., Crombie, B., Stacey, G. & Roberts, D.M. (1991) Plant Physiol. **95**, 222-227.

Wick, S.M. (1989) In: Calcium Binding Proteins, Vol II Biological Functions, Thompson, M.P. (ed.), CRC Press, pp 21-45.

Pathogen and pest resistance in endophyte-infected tall fescue

Kimberly D. Gwinn, Ernest C. Bernard and
Charles D. Pless

Department of Entomology and Plant Pathology, The University of Tennessee, Knoxville, TN 37901-1071, USA

Introduction

Tall fescue, a cool season grass grown on 15 million hectares in the United States, is a major source of livestock forage. Although tall fescue is nutritionally comparable to other forage grasses, animals raised on tall fescue pastures often have poor performance and exhibit a number of symptoms collectively known as fescue toxicosis (Bacon et al., 1986; Sanchez, 1987). Symptoms associated with fescue toxicosis include depressed weight gain, reduced reproduction capacity, and reduced lactation. Fescue toxicosis is estimated to cost cattle producers $600 million/year in unrealized profits (Hoveland, 1991). Horse and sheep production are also affected (Sanchez, 1987).

Fescue toxicosis is now known to be highly correlated with the presence of an endophytic fungus (*Acremonium coenophialum* Morgan-Jones and Gams) within tall fescue tissues (Crawford et al., 1989; McDonald et al., 1989; Read and Camp, 1986). Animals which relied on endophyte-infected (E+) plants for 15% or more of their diet exhibited symptoms of fescue toxicosis (McDonald et al., 1989). Symptoms include, but are not limited to, reduced weight gains, disrupted reproductive functions, reduced milk production, and increased vasoconstriction (Solomons et al. 1989; Schmidt et al. 1982). Although alkaloids have been implicated in fescue toxicosis of grazing animals (Belesky et al., 1989), chemical agents causing fescue toxicosis are not known. However, ergopeptide alkaloids produce symptoms similar to those of fescue toxicosis (Smith et al. 1974; Solomons et al., 1989; Woods et al., 1966).

Fescue endophyte

Bacon et al. (1977) isolated an endophytic fungus, *Epichloe typhina* (Fries) Tulasne, from tall fescue plants in pastures where grazing animals were exhibiting symptoms of fescue toxicosis. Morgan-Jones and Gams (1982) reclassified the endophyte as *Acremonium coenophialum*. The fungus has no known sexual stage, but asexual sporulation occurs *in vivo* and can be induced *in vitro* (Bacon, 1988). In nature, the fungus spends its entire life cycle within the plant and does not rely on sporulation for dissemination. The only known means of endophyte dissemination is via seed (Siegel et al., 1985; Welty et al. 1987).

Fungal hyphae have been found in all plant parts except the root and leaf blades (Hinton and Bacon, 1985). Highest concentrations of fungal hyphae were found in the leaf sheath region of the plant. During flowering, the fungus entered the inflorescence stem and grew within the intercellular spaces. Fungal hyphae entered the floret and then colonized the ovary during its early developmental stages. In the seed, the endophyte was sequestered between epidermal cells of the scutellum and starchy endosperm (Hinton and Bacon, 1985). In the dormant seed, the embryo was not infected; infection occurred only in the later stages of emergence. Although host cell morphology did not appear to be altered (Hinton and Bacon, 1985), comparison of E+ and E- clones indicated that changes in tissue morphology did occur (Arachevaleta et al., 1989). E+ plants had narrower and thicker leaf blades; these blades also had a more erect habit than leaf blades of E- plants, which tended to be more flexible. Air spaces within the leaf sheath mesophyll developed earlier and more extensively in E+ plants. Although leaf area was consistently higher in E+ plants, specific leaf weight was more dependent upon genotype than an endophyte status (Hill et al., 1989).

Yates et al. (1989) found several types of alkaloids associated with endophyte-infected tall fescue. These included ergot-like alkaloids (ergopeptines and ergonovine), phenethylamine (halostachine), ß-carbolines (harmine and norharmane), diazaphenanthrines (peramine), pyrrolopyrizineone (perloline), and saturated pyrrolizidines (loline, norloline and their derivatives). Some compounds are known to be produced by the endophyte (ergot-like alkaloids) (Bacon, 1988); some are of plant origin and are produced by the plant in response to the presence of the fungus (peramine and the saturated pyrrolizidine alkaloids).

Mutualistic relationships

Because presence of the endophytic fungus is highly correlated with fescue toxicosis, many animal producers have renovated pastures and planted endophyte-free (E-) varieties. Although this strategy has been lauded as a 'cure' for fescue toxicosis, problems in establishing and managing these pasture have decreased grower confidence in the merits of renovation. Tall fescue and its endophyte coexist in a mutualistic relationship (Siegel et al., 1987; Clay, 1989). The fungus is protected, disseminated, and nourished by the plant. E+ plants are more resistant to drought, pathogens, and herbivores than are E- plants.

Drought

A number of studies have documented the effects of endophyte infection upon the ability of tall fescue to withstand drought (Arachevaleta et al., 1989; West et al., 1988). Under drought conditions, pastures with an endophyte infestation level of 75% had significantly higher herbage yield and significantly lower percentage of dead tissue than did pastures with a zero infestation level (West et al., 1988). Regrowth of E+ plants was more rapid after exposure to drought conditions (Arachevaleta et al., 1989), and E+ plants were consistently more productive under drought stress. Pyrrolizidine and ergopeptine alkaloid synthesis was higher in drought conditions than under well-watered conditions (Belesky et al. 1989).

Pathogens

Conversely, little work has been published on the resistance of tall fescue to fungal pathogens. Crown rust (*Puccinia coronata* Corda.) was more severe on E- than on E+ fescue (Ford and Kirkpatrick, 1989); however, stem rust (*Puccinia graminis* Pers.) incidence and severity were independent of endophyte status of the plant (Welty and Barker, 1990). *A. coenophialum* was antagonistic to other fungi in culture (Siegel et al., 1987; White and Cole, 1985), but in some cases growth reduction was due to nutrient competition (Gwinn and Bernard, 1988). When leaves of tall fescue were inoculated with *Rhizoctonia zeae* Voorhes, severity or incidence of the resultant disease did not differ significantly between E+ and E- plants (Gwinn and Bernard, 1988). However, when seeds were planted in a soil mix infested with *R. zeae* or *R. solani* KHhn, percentage reduction in stand was negatively correlated with endophyte infestation level of the seed lot (Gwinn and Gavin, 1990; Blank et al., 1991).

Insects

Endophyte-infected grasses are resistant to insects, and endophytes are considered prime biocontrol agents (Clay, 1989). Endophyte-infected grasses are resistant to many foliage feeding insects, e.g. Argentine stem weevil, *Listronotus bonariensis* Kuschel (Barker et al., 1983); fall armyworm, *Spodoptera frugiperda* (J.E. Smith) (Clay et al., 1985, Hardy et al., 1986); oat-bird cherry aphid, *Rhopalisiphum maidis* (L.) (Ford and Kirkpatrick, 1989, Johnson et al., 1985, Latch et al., 1985); greenbug, *Schizaphis graminum* (Rondani) (Johnson et al., 1985); corn flea beetle, *Chaetocnema pulicaria* Melsheimer (Kirfman et al., 1986); billbugs, *Sphenophorus* sp. (Johnson-Cicalese and White 1990); and several species of leafhoppers (Kirfman et al., 1986).

Some insects are actively deterred from foliage of E+ tall fescue and many species of aphids cannot be forced to feed on these plants (Johnson et al., 1985). The chemical basis of feeding deterrence was shown to be due at least in part to the presence of saturated pyrrolizidine alkaloids (Johnson et al., 1985). Aphids did not survive on grasses containing loline alkaloids (Siegel et al. 1990). In a study on the effects of ergot alkaloids on feeding behavior and larval survival of southern armyworm *Spodoptera frugiperda*, J. E. Smith, Clay and Cheplick (1989) found that all types of ergot alkaloids (clavines, lysergic acids and ergopeptines) deterred feeding, but only ergonovine reduced larval survival. Ergovaline-containing grasses did not inhibit survival of aphids (siegal et al. 1990). Yates et al. (1989) estimated oral ED_{50} values for six alkaloids in milkweed bug *Oncopeltus fasciatus* (Dallas). N-formyl loline was more toxic than halostachine, norhamane or perroline; ergonovine and ergocryptine (an ergopeptine similar to ergovaline) had intermediate activities. A synergistic response occurred when insects ingested both perloline and ergonovine, but no other pairs showed synergistic activity. *Folsomia candida* Willem (Hexapoda: Collembola) showed no feeding response or population decline when fed yeast impregnated with ergot alkaloids or loline alkaloids (Bernard et al., 1990).

Resistance of endophyte-infected tall fescue to foliage feeding insects is well documented, but effects on insect stages that feed on plant roots have not been well studied. In a study on Japanese beetle, *Popillia japonica* Newman, Oliver et al. (1990) found that numbers of surviving Japanese beetle larvae were significantly ($p=0.05$) higher in the clay pots containing predominantly E- plants than in pots containing either the predominately E+ or all E+ plants. Survival in the pots containing only E- plants was significantly different from that in any other treatment. In a preliminary experiment, in which root zones of potted E+ and E- fescue plants were artificially infested with annual white grubs (*Cyclocephala* sp.) Oliver (1990) determined that survival of larvae was significantly higher on E- plants than on E+ plants. Cole et al. (1990) developed a bioassay in which larvae of *Drosophila melanogaster* Meigen were fed lyophilized tall fescue tissue mixed with commercial diet. Cumulative numbers of pupae and F_1 adults were counted each day for 18 days. On E+ leaf diet, few (= 1.4) *D. melanogaster* reached the pupal stage during the 18-day test period; on the E- leaf diet, a mean of 168 insects per rearing container pupated during the same period. Numbers of pupae on E+ and E- leaf diet were significantly different, beginning on day 6 and continuing until termination of the experiment. Survival to the pupal stage was also significantly lower on E+ root diet than on E- root diet beginning on day 10. Differences in larval survival on root diets were not as extreme as on leaf diets; however, by day 18, survival was more than five times greater on the E- root diet (= 78) than on the E+ root diet (= 15).

Nematodes

Information concerning the relationship of plant-parasitic nematodes to tall fescue is confused and partially contradictory before 1987, because the endophyte status of fescue was not recognized as a potential factor in the nematode-plant reaction. Although a number of ectoparasitic nematodes have been reported to be successfully reproducing parasites or pathogens of *F. arundinacea* (Hoveland et al., 1975; McGlohon et al., 1961; Pedersen and Rodriguez-Kabana, 1984, 1985), in few of these cases has the endophyte status been known. Ectoparasite densities, primarily those of *Helicotylenchus dihystera* (Cobb) Sher and *Paratrichodorus minor* (Colbran) Siddiqi, were reduced in pastures with high endophyte infestation levels (Pedersen et al., 1988; West et al., 1988). In greenhouse experiments, *H. pseudorobustus* (Steiner) Golden densities declined slightly on E+ fescue (Kimmons et al., 1990). West et al. (1988) found that *Tylenchorhynchus acutus* Allen densities under E+ fescue were only 25% of those under E- fescue. On the other hand, Halisky and Myers (1989) questioned whether *Acremonium* endophytes had any real effects on nematodes in turf soils, considering that growing conditions and sampling methods were more important determinants of nematode densities.

Effects of tall fescue on endoparasitic nematodes are more clearly understood, although the actual mechanisms resulting in plant resistance are not known. In greenhouse pot experiments, *Hoplolaimus galeatus* (Cobb) Thorne reproduced well on tall fescue cultivar 'Ky 31' (McGlohon et al., 1961), and Hoveland et al. (1975) reported it to be common in Alabama pastures. Reproduction of *H. galeatus* was twice as great on a fescue genotype with large-diameter roots as on a genotype with roots of smaller diameter (Rodriguez-Kabana et al., 1978). Unfortunately, the endophyte status of all of these genotypes was unknown.

Species of *Pratylenchus* are known to be suppressed by tall fescue, although endophyte status has not always been reported. Populations of *Pratylenchus penetrans* (Cobb) Filip. & Sch. Stekh. were reduced 20-40% after eight weeks on tall fescue of unknown endophyte status (Townshend et al., 1984). *P. scribneri* Steiner increased fivefold on tall fescue of unknown endophyte status (Minton, 1965), but *P. scribneri* populations in fescue pastures with high endophyte infestation levels were only 1% of their levels in E- pastures (West et al., 1988). This reduction was cited as a partial explanation for drought tolerance in endophyte-infected fescue. Kimmons et al. (1990), in greenhouse experiments, found that *P. scribneri* declined to non-detectable levels in the roots and rhizosphere of E+ plants, but maintained itself near initial inoculum levels on E- plants. In roots, nematode numbers were similar up to 29 days after infestation, but dropped to very low levels by day 58 on E+ root systems. Reproduction as determined by the presence of many small juveniles was easily seen in E- roots but was virtually absent in E+ roots (Bernard and Gwinn, 1991). Resistance of E+ tall fescue to nematodes appears to be due not to a single mechanism, but rather to a combination of chemical and morphological changes. Each of these changes influences individual interactions between nematode species and the grass.Little research has been done on parasitism of tall fescue by root-knot nematodes. Chapman (1973) reported the inability of *Meloidogyne hapla* Chitwood and *M. incognita* (Kofoid & White) Chitwood to parasitize seven tall fescue cultivars, but McGlohon et al. (1961) found that one isolate of *M. incognita* parasitized and reproduced on tall fescue, while another isolate did not. The endophyte status in these studies was not reported. A recently described species, *M. marylandi*, increased up to fivefold on E- fescue in greenhouse experiments, but did not increase on E+ fescue (Kimmons et al., 1990). Egg production on E+ fescue was less than 10% of that on E- fescue. When fescue and white clover were grown together, hatching and root invasion by an undescribed clover-parasitic *Meloidogyne* sp. were not affected (Kimmons et al., 1990), suggesting that either 1) a root-produced toxic or nemastatic exudate is highly specific to certain root-knot nematodes or is metabolized quickly by soil microflora, or 2) the resistance phenomenon is toxic only to nematodes not residing within the roots. We have also found that *M. marylandi* Jepson & Golden almost exclusively parasitizes larger-diameter roots in E- fescue (Bernard and Gwinn, 1991). After 13 to 17 days, hatching (juveniles in soil + roots) was higher in the E- treatments. Root invasion was consistently higher in E- root systems after day 6, and rose dramatically by day 17. These findings suggest that at least two separate phenomena may be involved in resistance exhibited by E+ fescue : 1) failure to exude a hatching stimulus; and/or 2) lack of attraction to or repellency from E+ roots.

The tall fescue-endophyte system provides an unique model for examination of mechanisms of resistance to pests and pathogens since genetically identical clones differing only in the presence of the endophyte are available. Also, degree of resistance differs among individual clones (unpublished data). Ability to produce ergot alkaloids *in vitro* also appears to vary among individual isolates tested (Bacon, 1988). It is likely, therefore, that E+ plants which do not produce mammalian toxicity factors but are resistant to abiotic and other biotic stresses are present within current populations. These mechanisms of resistance are poorly understood and elucidation of mechanisms may assist in the development of methods for selection.

References

Arachevaleta, M., Bacon, C.W. and Hoveland, C.S. 1989. Agron. J. 81:83-90.

Bacon, C.W. 1988. Appl. Environ. Microbiol. 54:2615-2618.

Bacon, C.W., Lyons, P.C., Porter, J.K. and Robbins, J.D. 1986. Agron. J. 78:106-116.

Bacon, C.W., Porter, J.K., Robbins, J.D. and Luttrell, E.S. 1977. Appl. Environ. Microbiol. 34:576-581.

Barker, G.M., Pottinger, R.P. and Addison, P.J. 1984. N. Z. J. Agric. Res. 27:279-281.

Belesky, D.P., Stringer, W.C. and Plattner, R.D. 1989. Ann. Bot. 64:343-349.

Belesky, D.P., Stuedemann, J.A., Plattner, R.D. and Wilkinson, S.R. 1989. Agron. J. 80:209-212.

Bernard, E.C., Cole, A.M., Oliver, J.B. and Gwinn, K.D. 1990. In: S.S. Quisenberry and R.E. Joost (ed.) Proc. Int. Symp. *Acremonium*/Grass Interactions, p. 125-127.

Bernard, E.C. and Gwinn, K.D. 1991. J. Nematol. (In Press).

Blank, C.A., Gwinn, K.D. and Gavin, A.M. 1991. Phytopathology (In Press).

Chapman, R.A. 1973. Univ. Ky. Agric. Exp. Sta., 85th Ann. Rept. p. 94.

Clay, K. 1989. Mycol. Res. 92:1-12.

Clay, K. and Cheplick, G.P. 1989. J. Chem. Ecol. 15:169-182.

Clay, K., Hardy, T.N. and Hammond, A.M., Jr. 1985. Oecologia (Berlin) 66:1-5.

Cole, A.M., Pless, C.D. and Gwinn, K.D. 1990. In: S.S. Quisenberry and R.E. Joost (ed.) Proc. Int. Symp. *Acremonium*/Grass Inter. p. 128-131.

Crawford, R.J., Jr., Forwood, J.R., Belyea, R.L. and Garner, G.B. 1989. J. Prod. Agric. 2:147-151.

Ford, V.L. and T.L. Kirkpatrick. 1989. In: C.P. West (ed). Proc. Ark. Fescue Toxicosis Conf. Ark. Exp. Sta. Univ. Ark. 1989. Special Report 140 p.

Gwinn, K.D. and Bernard, S.C. 1988. Phytopathology 78:1524.

Gwinn, K.D. and Gavin, A.M. 1990. Phytopathology 80:1052.

Halisky, P.M. and Myers, R.F. 1989. In: Rutgers Turfgrass Proceedings. 1989:124-131.

Hardy, T.N., Clay, K. and Hammond, A.M., Jr. 1986. Environ. Entomol. 15:1083-1089.

Hill, N.S., Stringer, W.C., Rottinghaus, G.E., Belesky, D.P., Parrot, W.A. and Pope, D.D. 1990. Crop Sci. 30:156-161.

Hinton, D.M. and Bacon C.W. 1985. Can. J. Bot. 63:35-42.

Hoveland, C.S. 1991. J. Agric. Ecosys. Environ. (In Review).

Hoveland, C.S., Rodriguez-Kabana, R. and Berry, C.D. 1975. Agronomy Journal 67:714-717.

Johnson, M.C., Dahlman, D.L., Siegel, M.R., et al. 1985. Appl. Environ. Microbiol. 49:568-571.

Johnson-Cicalese, J.M. and White, R.H. 1990. J. Am. Soc. Hort. Sci. 115:602-604.

Kimmons, C.A., Gwinn, K.D. and Bernard, E.C. 1990. Plant Disease 74:757-61.

Kirfman, G.W., Brandenburg, R.L. and Garner, G.B. 1986. J. Kansas Entom. Soc. 59:552-554.

Latch, G.C.M., Christensen, M.J. and Gaynor, D.L. 1985. N. Z. J. Agric. Res. 28:129-132.

McDonald, W.T., McLaren, J.B., Chestnut, A.B., Fribourg, H.A., Keltner, D.G. and Carlisle, R.J. 1989. J. An. Sci. 67:47.

McGlohon, N.E., Sasser, J.N. and Sherwood, R.T. 1961. N.C. State Univ. Exp. Sta. Tech. Bull. 148:1-39.

Minton, N.A. 1965. Plant Dis. Rep. 49:856-859.

Morgan-Jones, G. and Gams, W. 1982. Mycotaxon 15:311-318.

Oliver, J.G., Pless, C.D. and Gwinn, K.D. 1990. In: S.S. Quisenberry and R.E. Joost (ed.) Proc. Int. Symp *Acremonium*/Grass Interactions, p. 173-175.

Pedersen, J.G., Rodriquez-Kabana, R. and Shelbey, R.A. 1988. Agron. J. 80:811-814.

Read, J.C. and Camp, B.J. 1986. Agron. J. 78:848-850.

Rodriquez-Kabana, R., Haaland, R.L., Elkins, C.B. and Hoveland, C.S. 1978. J. Nematol. 10:279.

Sanchez, D. 1987. Agric. Res. 35:12-13.

Schmidt, S.P., Hoveland, C.S., Clark, E.M., Davis, N.D., Smith, L.A., Grimes, H.W. and Holliman, J.L. 1982. J. An. Sci. 55:1259-1265.

Siegel, M.R., Latch, G.C.M., Bush, L.P., Fannin, F.F., Rowan, D.D., Tapper, B.A., Bacon, C.W. and Johnson, M.C. 1990. J. Chem. Ecol. 16:3301-3315.

Siegel, M.R., Latch, G.C.M. and Johnson, M.C. 1985. Plant Disease 69:179-183.

Siegel, M.R., Latch, G.C.M. and Johnson, M.C. 1987. Ann. Rev. Phytopath. 25:239-315.

Smith, V.G., Beck, T.W., Convey, E.M. and Tucker, H.A. 1974. Neuroendocrinology 15:172-181.

Solomons, R.N., Oliver, J.W. and Linnabary, R.D. 1989. Amer. J. Veter. Res. 50:235-238.

Townshend, J.L., Cline, R.A., Dirks, V.A. and Marks, C.F. 1984. Can. J. Plant Sci. 64:355-360.

Welty, R.E. Azevedo, M.D. and Cooper, T.M. 1987. Phytopathology 77:893-900.

Welty, R.E. and Barker, R.E. 1990. Phytopathology 80:1043.

West, C., Izekor, E., Oosterhuis, D.M. and Robbins, R.T. 1988. Plant and Soil 112:3-6.

White, J.F., Jr. and Cole, G.T. 1985. Mycologia 77:487-489.

Woods, A.J. Jones, J.B. and Mantle, P.G. 1966. Vet. Res. 78:742-749.

Yates, S.G., Fenster, J.C. and Bartelt, R.J. 1989. J. Agric. Food Chem. 37:354-357.

The commercial pathway for agricultural biotechnology

Zachary S. Wochok

Calgene, Inc. 1920 Fifth Street, Davis, CA 95616, USA

Introduction

Considerable progress has been made over the past decade in the area of agricultural biotechnology. From a position of science development in the 1970's, this field has progressed through a period of technology development to one of product development and pre-commercialization. As we enter the last decade of this millennium, we embark on a period during which genetically engineered crops will be introduced into commercial agriculture.

In order to assess the current situation and the outlook for the future, it is useful to review the factors and strategies that have led to the current levels of commercialization. This is best accomplished by reviewing the history of agricultural biotechnology companies over the past decade. For the purpose of providing focus to the discussion of commercialization, a case study is presented herein which is drawn from experiences at Calgene, Inc.

Key Ingredients

Agricultural biotechnology companies in the private sector are coming of age as they move from the development stage to fully operational businesses. In 1990, the average age of the major publicly-held companies was about ten years. Within that relatively brief period, some companies merged their operations to gain improved efficiencies of scale. Most notable among these was the merger of Advanced Genetic Sciences with DNA Plant Technologies, and Plant Genetics, Inc. with Calgene, Inc. In 1990 the seven publicly traded U.S. agricultural biotechnology companies collectively generated revenues of about 55 million dollars. Based on projections available through publicly available information, it is reasonable to expect that these companies, or combinations of them, should be generating three to five times that amount of product revenues by the mid-1990's.

As agricultural biotechnology companies move towards developing and selling more products, particularly products from advanced technologies, what are the key ingredients that will determine the eventual success of their efforts? As one looks at the relative successes of companies today, several key ingredients become apparent. Successful companies all have a well-defined goal and a strategic plan for achieving that goal. Executing the plan requires an experienced management team as there is little margin for error for relatively young entrepreneurial companies. The business team must be sensitive to the marketplace and assist the technology team in identifying those targets having the maximum value to the end user. In addition they must be capable of managing the delicate balance of technology push and market pull that exists in any technology driven organization. Thus, in addition to providing marketing insights directly, they must be highly responsive to new technology developments emanating from either their own or other laboratories, be able to assess their commercial value, determine whether they fit the overall corporate goal, and take appropriate action. The ability of a smaller, highly focused company to respond rapidly to such conditions and situations gives them a distinct competitive edge over larger, more bureaucratic organizations. In the final analysis, there is room for both. End users of products developed from these advanced technologies will be best served by having both small and large companies cooperate to the benefit of those who will ultimately be the recipients and benefactors of these technologies: the farmer, the distributor, the food processor, the retailer, and the general consumer.

Technoeconomic Dynamics

To fully appreciate the contributions that can be derived from biotechnology in agriculture, it is necessary to understand the economic realities surrounding the world agricultural marketplace. It soon becomes apparent that new technologies must be driven in the direction of those markets where economics dictate the need for certain improvements. The question then becomes whether a given technology can in fact provide those improvements. It is this convergence of technological and economic forces which I have termed technoeconomic dynamics.

The technoeconomic dynamic model shown in the Figure 1 demonstrates that where the market forces exert downward pressure on pricing, the net returns to the producer can be enhanced by bringing about improvements which result in lower production costs. In those cases where there is a clear market demand, the producer has a double opportunity to enhance operating performance by lowering costs and obtaining a higher price for the product, where that is justified, by providing greater value to the user, a portion of which is shared by each. Clear examples of these in agriculture can be seen with corn, tomato, and other hybrids, which command a higher price based on the value received in improved field performance. Biotechnology companies recognizing those kinds of opportunities, and having the ability to use their technologies to address those needs, will indeed be successful.

One of the key elements related to the success of a business venture of any kind is financial strength. With biotechnology companies, this is an even more critical factor given the long lead times required to develop and commercialize any agricultural product, let alone one derived from genetic engineering. For this reason it is important to have the ability to rapidly transfer technology know-how to product evaluation and development. To accomplish this requires a strong commercial development capability among staff members, and finally, if not most

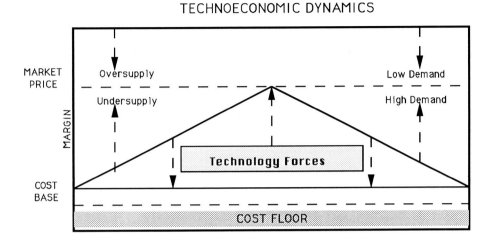

Figure 1. The technoeconomic dynamics model demonstrates the relationship between market supply and demand, production costs, and technology. Introduction of new technology can add sufficient value to alter supply/demand dynamics and/or reduce production costs resulting in improved profitability.

importantly, having the resources required to take a product to full commercialization.

Commercial Opportunities

There are two primary opportunities for agricultural biotechnology companies: microbial-based biological pest control products, and crop improvement resulting in seed or food and feed products.

Several agricultural biotechnology companies have successfully marketed products developed through plant breeding, cell culture and microbial selection. These are shown in Table 1. While there are no genetically engineered products as yet which have been developed through recombinant DNA technology, several are now in advanced stages of development and undergoing extensive field trials. Based on publicly available information, a new generation of products is expected by the end of this decade. Some of these are shown in Table 2.

Microbial based products are being sold today under different labels. These are primarily based on *Bacillus thuringiensis*. Other microorganisms will be used to develop a variety of other pest control products in the future. Among the first genetically engineered crops expected to be commercialized are cotton, tomato, potato and rapeseed.

Table 1. Partial product listing of major U.S. publicly-held agricultural biotechnology companies in the past seven years.

Proprietary Products Sold by Agricultural Biotechnology Companies

Company	Products	Year Introduced
Calgene	Nu-Spud™ Seed Potato Tubers	1984
	PGI™ Alfalfas	1985
	PGI™ Hybrid Tomatoes	1985
	Stoneville Cotton	1986
	Ameri-Can Canola Seed	1988
	Cano-Lite Canola Oil	1990
	NobleBear Corn	1990
	Erucical TD-13	1990
DNAP	Snow-Max™	1986
Ecogen	Dagger G™	1988
Mycogen, Inc.	M-One™	1988

The benefits of these new technologies can be seen in several ways. First of all, they provide a more focused approach to product development because they allow precise isolation of genetic traits which can be transferred directly to a crop or organism of choice. The commercialization time, while still lengthy by virtue of the limitations associated with plant growth and development cycles, will be shorter compared with more traditional technologies. Further, genetic engineering, like hybridization before it, provides a process for developing proprietary products with higher value, and a greater potential to recapture the investment made in developing these products.

These benefits can be seen upon evaluating both product opportunities. Biological pesticides can be produced in a relatively shorter time than traditional chemical pesticides. They also provide a valuable alternative to those organic chemicals now being used, or those which have been taken out of commerce. Their rapid biodegradability provides an environmental incentive toward adoption in pest management programs in concert with, or as an alternative to, chemical control.

In the area of crop improvement through genetic engineering of plants, the same benefits hold. A comparison of traditional breeding programs and a revision of these programs through the implementation of recombinant techniques shows the time reduction benefit. In a traditional breeding program, variety development can take up to eight years. Recombinant technology reduces this time dramatically by providing a mechanism for going directly to limited field trials followed by a replicated field trial and then commercialization. The one constant in the equation is the need for field trials to demonstrate agronomic efficacy.

Table 2. *Products of major U.S. publicly held agricultural biotechnology companies expected to be marketed in the 1990's.*

Products Expected by 1995

Company	Products
Biotechnica, Int'l	Rhizobium Seed Treatment
	Alfalfa, Soybeans, Corn Hybrids, and Wheat Seeds
Calgene, Inc.	Modified Rapeseed and Canola Oil
	Flavr-Savr™ Tomato
	Bromoxynil and Insect Tolerant Cotton
	PGI™ Corky Root Rot Tolerant Lettuce
	PGI™ Virus Tolerant Potatoes
	PGI™ Alfalfas
DNAP	Vegi-Snax
Ecogen	Foil™ Potato Biopesticide
	Condor™ Hardwood Biopesticide
	Cutlass™ Garden Biopesticide
ESCAGenetics	Date Palm Plantlets, Phyto Vanilla
	True Potato Seed, Sweet Corn
Mycogen Corporation	MCap™ Bt Bioinsecticide
	Casst™ Bioherbicide
	MYX 1200 & 1621 Bioherbicide

While the regulatory issues regarding these products are still subjects of discussion at local, state, and federal levels, it appears that the movement is toward a generally positive approach towards the development and use of biological pesticides and genetically engineered crops using various biological technologies, including genetic engineering.

A good way to evaluate the progress that has been made in the field of agricultural biotechnology is to consider an actual case study.

Calgene Case Study

With the foregoing status report of the U.S. agricultural biotechnology industry, the remainder of this paper will focus on some specific examples of how genetically engineered products are being developed at Calgene and the company's status in the commercialization process.

Calgene, Inc. was founded in 1980 as a technology based company. As technology development proceeded, the company made the transition to a more fully integrated operating company. An important part of this transition has been the focused approach Calgene has taken towards product development. The primary objective of the company is to develop improved plant varieties and plant products for the seed, food, and specialty chemical industries. The strategy of the company is equally well defined, and that is, "to be a leader in genetic engineering, to address major crop markets, develop proprietary products, while operating profitable businesses and working with international partners to gain maximum return on dollar invested." The choice of product targets has been based on the market opportunity, an analysis of the competition in the product area, and the applicability of the technology to achieve the targeted improvements within a reasonable period of time. As a result, the company has focused on several proprietary, high value products. These are: 1) herbicide- and insect-tolerant cotton, 2) extended shelf-life and more flavorful tomatoes, 3) superior industrial and edible vegetable oils, and 4) agronomically and qualitatively improved potatoes for planting stock and processing.

One of the guiding principles for Calgene has been the establishment of operating businesses in those areas where the company planned to utilize its technology. As a result, in 1991 Calgene was operating five separate subsidiaries and two joint ventures, all of which were producing, marketing and selling products in the core crop areas dictated by the corporate strategic plan. Technology programs were directed towards specific targets applicable to the company's major crops. The strength of the company's technology is evidenced by its numerous scientific publications, and over 40 issued patents generated by company scientists. The level of progress in transferring this technical progress to commercial reality is evidenced by the level of commitment to product development, best measured by evaluating the product introduction timelines of genetically engineered crops as shown in Figure 2.

One additional element in the commercialization of genetic engineering technology in crops is the regulatory aspect. Field trials of genetically engineered plants are regulated by the U.S. Department of Agriculture. Prior to being released to commerce, all genetically engineered crops used as food directly, such as fresh tomatoes, or indirectly as animal feed, such as cotton meal, are subject to certain regulatory guidelines already established by the Food and Drug Administration for foods in general. While there has been controversy relating to the perceived need for special regulations related to genetically engineered crops, the agency already has in place a very intricate and detailed set of regulations that are well suited to evaluating the safety and efficacy of foods irrespective of the technology used to produce the food. This has heretofore represented an as yet untested area of the commercialization pathway for genetically engineered crops. In the latter part of 1990, Calgene was the first company to submit a petition to the FDA for approval of the *kanamycin* marker gene used in the process of introducing specific gene constructs into plants. The evaluation process may take as long as two years but may be completed in a shorter period depending on the type of follow-up data that may be required.

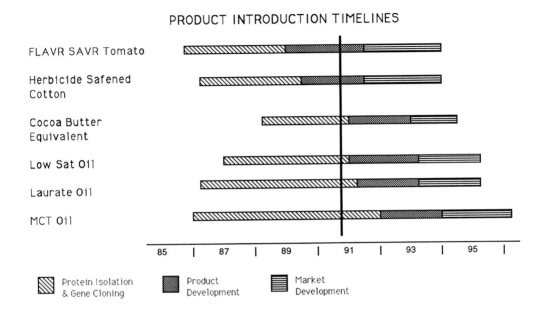

PRODUCT INTRODUCTION TIMELINES

Figure 2. Timelines for Calgene's genetically engineered products show level of progress in transferring technology to estimated commercial release. Timelines are subject to change based on milestone results.

One of Calgene's core crops is cotton. There are several reasons why this particular crop was selected for genetic engineering. Cotton is the fifth largest U.S. crop, planted on 10 million acres. The four billion dollar cotton fiber market is growing versus petroleum based synthetic fibers. Cotton generates high returns for growers. Dramatically lower costs and improved yields can be achieved by providing improved pest control in seed through genetic engineering. Finally, there are few seed competitors and the business provides attractive margins.

Herbicide safened and insect resistant cotton were chosen as the targets for genetic engineering based on economic evaluations. For example, U.S. cotton growers spend 150 million dollars annually on chemicals to control weeds. The most effective weed control chemicals are not currently used on cotton because they damage the crop. Bromoxynil, a low dosage, environmentally sound herbicide can be used with safened cotton varieties for lower cost and improved yields.

Cotton growers in the U.S. spend between 200 to 350 million dollars annually to control cotton insect pests. Insect resistance allows the plant to produce its own bioinsecticide thereby achieving improved insect control with reduced chemical costs.

In 1990 Calgene conducted two field trials of its genetically engineered cotton, and 34 field trial sites are planned for 1991. Calgene has genetically engineered commercial lines of cotton with its patented bromoxynil safened gene alone and in combination with a *Bacillus thuringiensis* gene construct for insect resistance. Bromoxynil is manufactured by Rhône-Poulenc, a research and development partner of Calgene which has supported the company's cotton research efforts for the past several years. Bromoxynil is an attractive herbicide target because of its broad weed control spectrum and its brief half-life in the environment, breaking down in the soil within one to two weeks after being applied. Development of bromoxynil safened cotton will permit growers to topically apply bromoxynil on mature cotton stands to combat weed infestation without damaging the cotton. The benefit to Rhône-Poulenc is that they can now sell their herbicide into a new market. For Calgene the benefit is that its subsidiary company, Stoneville Pedigreed Seed Company, will be able to sell these new proprietary high value cotton varieties in their markets. The added value is derived from the considerable savings afforded cotton growers in the form of improved field performance. From an environmental perspective this approach is attractive since it provides an alternative weed control treatment that can replace existing chemicals which have longer soil persistence. A further benefit is that by replacing several chemicals requiring relatively high application rates with one requiring relatively low rates, total chemical usage should be substantially reduced without compromising weed control. Calgene's business relationship with Rhône-Poulenc exemplifies how the choice of a strong partner with convergent interests, appropriate target selection, and the availability of a direct system of product distribution fits the attributes of key ingredients described earlier.

The Bt strategy for cotton is also straightforward in terms of providing an alternative to chemical insect control. Its value is measured by the savings attributed to intrinsic pest control versus the recognized costs associated with chemical pest control. It is also consistent with other criteria established for selecting appropriate gene targets. Calgene's petition to the FDA for the use of the *kanamycin* gene marker covered cotton, tomato and rapeseed. The company expects to follow with specific gene petitions within the next year. Based upon the length of the approval process, Stoneville Pedigreed Seed Company should be expected to sell genetically engineered cottonseed by the mid 1990's.

In summary, successful commercialization of any new agricultural product is never a given -- regardless of the product's degree of innovation, superior performance or advanced technology. Only with a clear, well-defined strategic plan at the outset can the commercial pathway be travelled at all. Integration of technology, product, and market development must be carried out with much precision in order to open the door to the marketplace. In the final analysis, it is the consumer who ultimately decides. It is a path oft taken, a path well worn. We in agricultural biotechnology are but its most recent traveller.

Glossary

The following terms are defined in a general sense. Specific details are avoided to allow the general reader and the developing student to become literate in the language (and jargon) underlying molecular biology and genetics.

AFLP: amplification fragment length polymorphism. A variant DNA amplification product produced by DAF, PCR, or RAPD of different size.

Agrobacterium rhizogenes: bacterium causative for the hairy root disease, a form of organized tumor similar to the crown gall disease. Tumors also contain bacterial DNA transferred and covalently integrated in the plant genome.

Agrobacterium tumefaciens: bacterium causative for the crown gall disease. capable of tumor induction in plants. This is achieved through the transfer of a region of DNA (T-DNA) which codes for the synthesis of plant growth regulators.

allele: one of the forms of a gene

aneuploid: chromosomal number is more or less than the diploid number by sets lesser than the haploid number (eg. a humans with 47 or 45 chromosomes are aneuploids.

antibiotic: substance which acts to destroy or inhibit the growth of a microbe (eg. bacteria or fungi).

antibody: common name for an immunoglobulin protein molecule which reacts with a specific antigen.

antigen: foreign molecule recognized by the immune response system of an animal.

ATP: adenosine triphosphate. The general energy donor molecule in all organisms.

autoradiography: method used to detect radioactive substances by the property to darken film superimposed on the compounds. Can be used on whole organisms or molecules separated by molecular methods such as electrophoresis.

auxin: plant hormone involved in cell elongation and growth (eg. indole acetic acid, 2,4-D).

bacteroid: the nitrogen-fixing form of a symbiosome-contained *Rhizobium* or *Bradyrhizobium* bacterium.

biolistics: process by which DNA molecules are propelled into a recipient cell using coated microprojectiles shot from a 'gene gun'. The method of propulsion may vary and ranges from electric discharge to helium blast.

biotechnology: the combination of biochemistry, genetics, microbiology, and engineering to develop products and organisms of commercial value.

Bradyrhizobium: bacteria able to form nodules and fix nitrogen in association with some legumes such as cowpea, soybean and peanuts.

cDNA: complementary DNA; DNA made by reverse transcriptase enzyme from RNA.

centimorgan: cM; unit of recombination equal to 1 percent recombination.

centromere: chromosomal region functioning as the spindle attachment region to allow chromosome and/or chromatid separation during mitosis and meiosis.

Chargaff's rules: stipulate that since in double stranded DNA the amount of adenine (A) equals that of thymine (T) and guanine (G) that of cytidine (C). Accordingly A binds to T and G to C by hydrogen bonds, giving the DNA molecule the properties needed for replication and information storage.

chloroplast: site of photosynthesis in eukaryotes. Contains circular DNA.

chromatin: complex form of eukaryotic nuclear material at the times between cellular divisions.

chromosome walking: strategy of chromosome analysis in which cloned chromosomal segments are used to isolate the neighboring DNA fragments.

chromosome: organized structure made up of DNA and proteins. Visible through light microscopy during cell division (cf. mitosis and meiosis).

clone: based on the Greek word 'klon' meaning 'twig'. A method of vegetative reproduction of an organism. Commonly used in horticulture as cuttings or drafts. Resulting organisms are defined as a clone meaning they were derived from the same original source organism. Commonly clones are presumed to be genetically identical. This may not be the case because of further genetic change after the original duplications. In modern genetic jargon, clone describes an isolated DNA sequenced ligated into a bacterial plasmid or virus, so that the sequence can be propagated indefinitely using microbiological means. Such cloned sequence can be used for sequencing, expression studies, or as probe.

codon: arrangement of three nucleotides in mRNA controlling the insertion of an amino acid into a polypeptide.

contig: contiguous fragment of DNA used in genome analysis.

cortex: bulk tissue of a plant root or legume nodule. Characterized by vacuolated cells and absence of mitotic divisions.

cytokinin: plant hormone involved in cell division and senescence (eg. kinetin, zeatin).

DAF: DNA amplification fingerprinting. A method of general DNA amplification (cf. PCR) using a single primer of between 6 and 8 nucleotides in length. The primer is arbitrarily chosen and may generate as many as 80 amplification products, which can be resolved by a variety of methods, including polyacrylamide gel electrophoresis (PAGE) and silver staining, agarose electrophoresis, and automated analysis using either DNA sequencers (using fluorescent primers and laser detection) or capillary chromatography. Method was developed in 1990 by the University of Tennessee and is used to distinguish organisms (eg. cultivar of one crop species), gene mapping and diagnostics.

differentiation: process of cell and tissue specialization involving differential gene expression.

diploid: the normal somatic chromosome number of an organism (twice haploid).

DNA: deoxyribonucleic acid. The genetic molecule of most organisms, except some viruses. Double stranded polymer of nucleotides arranged along a deoxyribose and phosphate backbone. Structure was proposed by James Watson and Francis Crick in 1953.

electrophoresis: method used to separate protein or nucleic acid molecules in an electric field extending across a physical medium such as agarose gel or polyacrylamide. Derived from Greek: phoro, meaning I carry.

endodermis: cell layer separating cortex and pericycle in plant roots. Contains the Casperian Strip, which is important in diffusion resistance and turgor relationships in plants.

enzyme: a biological catalyst allowing the completion of biochemical reactions. Most enzymes are proteins although some RNA enzymes were recently discovered (see ribozyme).

epidermis: external cell layer of a plant.

Escherichia coli: also *E. coli*. A common gut bacterium used as a model genetic organism. *E. coli* has about 3000 genes and a genome of around 4 million basepairs.

ethidium bromide: chemical used to visualize DNA by fluorescence. Interpositions itself into the DNA groove and alters buoyancy.

ethylene: gaseous plant hormone involved in stress responses, fruit ripening as well as nodulation of legumes.

eukaryote: organism characterized by the presence of a nucleus. Also other organelles such as mitochondria and/or chloroplasts may be present in eukaryotes. Includes all plants, animals, green algae and fungi.

exon: expressed region of a gene. Transcribed and translated.

flavone: aromatic molecule (ie. contains a benzene ring as a core molecule) significant in the communication of legume plant to *Rhizobium* and *Bradyrhizobium*.

Frankia: an actinomycetes bacterium capable of forming nodules with some non-legumes such as Alnus and Casuarina (tree species).

gene: functional unit of inheritance. Usually a gene is defined as that region of DNA that controls the synthesis of a polypeptide.

genetic code: the conversion table which allows the interpretation of triplet codons to their matching amino acids.

genetic engineering: the directed genetic manipulation of an organism using recombinant DNA molecules not commonly found in nature.

genome: the entire set of hereditary molecules in an organism.

genotype: the genetic make-up of an organism which depending on the environment and other genes may be expressed to the phenotype.

haploid: the gametic chromosome number of an organism.

homogenotization: bacterial genetics procedure used to exchange genetic markers from a plasmid to the recipient linkage group. Also called marker exchange.

hormone: regulatory substance which acts at low concentrations (less than one micromolar) and at a distance from its site of synthesis. Controls metabolism and development.

intron: intervening sequence in genes. Transcribed but not translated.

isoflavone: aromatic signal substance involved in the nodulation of legumes.

karyotype: the pattern and shape of chromosomes of an organism.

kilobase: kb; one thousand basepairs of DNA.

lectin: protein molecule with no known function other than sugar-binding ability.

legume: plant family characterized by a pea like flower morphology. Many but not all legumes are nodulated and form nitrogen-fixing symbioses with soil bacteria called *Rhizobium, Bradyrhizobium,* and *Azorhizobium.*

locus: the chromosomal position of a genetic condition as defined by a detectable phenotype

map: the ordered arrangement of genes or molecular markers of an organism, indicating the position and distance between the markers and loci. Most maps are genetic maps based on the percentage recombination. Some maps are cytological maps based on the arrangement of chromosomal regions, while others are physical maps based on the amount of DNA between markers and loci.

megabase: Mb; one million basepairs of DNA.

meiosis: cell division in eukaryotes giving rise to gametes of different genetic make-up to the parental cell and to each other.

meristem: organized zone of mitotic division giving rise to cell clusters capable of further differentiation into new organ types.

messenger RNA: mRNA; product from DNA by transcription which serves as the information carrier for translation in proteins.

mitochondrion: organelle found in all eukaryotes. Site of respiration (ATP synthesis). Contains its own DNA.

mitosis: cell division in eukaryotes giving rise to two genetical identical progeny cells. Occurs frequently in meristems.

molecular marker: a molecular signpost used in eukaryotic gene isolation. Usually a RFLP probe or a primer site for DNA amplification.

mutant: an organism or gene with inheritable altered phenotype from the wild type.

NAR: nodulation in the absence of *Rhizobium*. Genetic condition found in alfalfa plants leading to the formation of nodules without *Rhizobium* being present.

nitrogen fixation: process by which nitrogen gas is converted to ammonia. This process occurs frequently in bacterial induced nodules of legumes and results in an independence on fertilizer nitrogen. Also possible by the industrial Haber-Bosch process.

nod-box: DNA sequence found in front of several gene groupings involved in the nodulation ability of *Rhizobium* and *Bradyrhizobium*.

nodulation factor: lipo-oligosaccharide molecule (containing sugars and a fatty acid) synthesized by *Rhizobium* or *Bradyrhizobium* in response to a plant signal (usually a flavone or isoflavone) capable of inducing cell division and root hair curling in legume roots.

nodule: outgrowth from the roots (or stems in some cases) of legumes induced by bacteria or exogenous agents such as bacterial derived nodulation factors or auxin transport inhibitors.

nucleotide: component of DNA and RNA. Two nucleotides paired according to Chargaff's rules are one base pair.

oligonucleotide: a polymer of nucleotides usually 5 to 30 base pairs long.

organelle: membrane bound cellular compartment (eg. nucleus, chloroplast, mitochondrion, Golgi apparatus).

PCR: Polymerase Chain Reaction. A method for amplifying DNA of any organism using two specific oligonucleotide primers (about 15 base pairs in length) which flank the region of interest. The method was developed by CETUS Corporation in the mid-1980s and is of extreme value in diagnostics, forensics and general molecular biology (e. g. sequencing, probe preparation, genome mapping).

peribacteroid unit: see symbiosome.

pericycle: cell layer surrounding the vascular bundle in plant roots. Gives rise to lateral roots as well as nodules in actinorhizal plants such as Alnus. Is also involved in legume nodule formation.

PFGE: see pulse field gel electrophoresis.

phenotype: the appearance of an organism taken as a genetic characteristic.

phosphorylation: biological process by which proteins are 'decorated' with phosphate groups derived from ATP. The process alters the biological activity of the protein whereby facilitating a form of physiological regulation.

phytoalexin: substance involved in the antimicrobial response of a plant.

plant growth regulator: broad class of chemicals which control the growth of plants. Many are also natural compounds found within plants, where they may act as hormones.

plasmid: circular, covalently closed DNA molecule commonly found in bacteria. Often used as a cloning vector in genetic engineering.

polyploid: more than diploid by multiples of the haploid number.

polysaccharide: polymer molecule of sugars (eg. starch, glycogen).

positional cloning: experimental approach used to locate and isolate gene sequences for which the gene product is not known. Instead the phenotype is mapped and large fragments are isolated in the region of informative molecular markers known to segregate closely with the gene of interest.

primer: short sequence of DNA (or RNA) used to initiate DNA replication.

probe: a known sequence of DNA (or RNA) used to detect homologous sequences in DNA or RNA after reassociation based on Chargaff's rules.

prokaryote: bacteria. Characterized by the absense of major organelles such as the nucleus and plastids.

promoter: regulatory region of a gene involved in the control of RNA polymerase binding to the target gene.

protein: a polymer of amino acids usually with structural roles (such as keratin, the hair protein) or catalytic roles (see enzyme).

pulse field gel electrophoresis: PFGE; a variation on the electrophoresis procedure in as far that a computer flips the electric field in preset pulses in different directions and defined strengths. Method is used to isolate very large DNA molecules (greater than one megabase; million basepairs) including whole chromosomes of organisms such as yeast.

RAPD: Random amplified polymorphic DNA. DNA amplification method similar to DAF, as it uses single primers. Method is restricted to primers 9 nucleotides and larger, and amplification products are generally visualized using agarose electrophoresis and ethidium bromide fluorescence. Usually about 4 to 10 products are generated. Method is useful for mapping approaches.

rDNA: recombinant DNA; made by the joining of DNA fragments from different species using restriction endonuclease and cloning approaches.

receptor: protein molecule usually on the cell able to receive and interpret an external signal.

recombination: natural process of exchanging DNA fragments between different DNA molecules. Occurs in both prokaryotes and eukaryotes, but by slightly different processes. Eukaryotic recombination occurs predominantly during meiosis and gives rise to gametes of non-parental gene combinations.

restriction endonuclease: an enzyme which cuts (or restricts) DNA at specific sequences generating fragments (eg. *Eco*RI).

reverse transcriptase: enzyme able to synthesize DNA from RNA. Often found in tumor viruses.

RFLP: restriction fragment length polymorphism. DNA fragment difference generated by the action of a restriction endonuclease on the DNA of two or more organisms and detected usually by Southern hybridization using a radioactively labeled probe.

Rhizobium: bacteria able to form nodules with some legumes such as peas, alfalfa, and clovers.

ribosome: site of protein synthesis in prokaryotes and eukaryotes. Made up of two subunits, comprising three RNA molecules and about 50 proteins.

ribozyme: enzyme made entirely of RNA.

RNA: ribonucleic acid. Used as messenger RNA, transfer RNA and ribosomal RNA. Some RNA molecules are the genetic molecule of viruses (eg. tobacco mosaic virus and HIV) and viroids. Some RNA molecules may have enzymatic activity (see ribozymes).

root hair: protruding from an epidermal cell. Grows by tip elongation.

Southern hybridization: also Southern blotting. Method employing gel separation of restricted DNA fragments, their blotting onto a membrane support, dissociation into single stranded DNA and hybridization (reassociation) with a labeled probe. Regions of homology are detected usually by autoradiography. Invented by Dr. Southern in 1975.

soybean: *Glycine max* (L) Merr. A tropical legume of widw agronomic application. Nodulates with *Bradyrhizobium japonicum* and *Rhizobium fredii*.

Stoffel fragment: truncated *Taq* polymerase without its exonuclease activity. Used in PCR and DAF. Named after the investigator Dr. Stoffel.

STS: sequence tagged site. Region of DNA on a chromosome used as a signpost for molecular gene mapping approaches.

supernodulation: ability of legume plants to form significantly higher nodule numbers than the normal. Found in several legumes such as pea and soybean.

symbiosis: mutually beneficial living together of two organisms of different species (eg. nodulation in legumes).

symbiosome: organelle structure found in nodules of legumes. Encases the symbiotic form of *Rhizobium* and *Bradyrhizobium* called the bacteroid. New term for 'peribacteroid unit (PBU)'.

Taq **polymerase**: thermostable DNA polymerase from *Thermus aquaticus.*

telomere: terminal region of chromosomes characterized by repeated DNA sequences.

Thermus aquaticus: thermophylic bacterium found in hot springs. Its DNA polymerase enzyme is thermostable and is used in the PCR, RAPD, and DAF.

transcription: the process by which DNA is copied into RNA. As the nucleic acid 'language' stays the same (see genetic code), the process is called transcription (cf. translation).

translation: the process by which RNA is made into proteins. Occurs in ribosomes. Called translation because the nucleic acid 'language' based on a sequence of four 'letters' arranged in triplet codons is changed to a 20 component 'language' used in protein synthesis.

wild type: the normal form of an organism (ie. not mutant). Note: this is a noun (not hyphenated).

wild-type: the normal, non-mutant form of an organism or allele. Note: this is the adjective and is hyphenated.

YAC: yeast artificial chromosome. Used extensively in the cloning of large DNA molecules used in eukaryotic gene mapping.

Index